Elementary Behaviour of Composite Steel and Concrete Structural Members

Elementary Behaviour of Composite Steel and Concrete Structural Members

Deric J. Oehlers
Department of Civil and Environmental Engineering
The University of Adelaide, Australia

Mark A. Bradford
Professor of Civil Engineering
School of Civil and Environmental Engineering
The University of New South Wales, Australia

CRC Press
Taylor & Francis Group
Boca Raton London New York

CRC Press is an imprint of the
Taylor & Francis Group, an **informa** business

A SPON PRESS BOOK

First published 1999

This edition published 2012 by Spon Press
2 Park Square, Milton Park, Abingdon, Oxon OX14 4RN
711 Third Avenue, New York, NY 10017, USA

Spon Press is an imprint of the Taylor & Francis Group, an informa business

British Library Cataloguing in Publication Data
A catalogue record for this book is available from the British Library

ISBN - 978 0 7506 3269 0

Contents

Preface	**ix**
Notation	**xi**
1 Introduction	**1**
1.1 Composite structures	**1**
1.2 Design criteria	**4**
1.2.1 General	4
1.2.2 Characteristic values	5
1.2.3 Limit states design	5
1.3 Material properties	**7**
1.3.1 General	7
1.3.2 Structural steel	7
1.3.3 Profiled steel	7
1.3.4 Reinforcing steel	8
1.3.5 Concrete	8
1.3.6 Rigid plastic properties	10
1.4 Partial shear connection	**12**
1.4.1 General	12
1.4.2 Equilibrium of forces	12
1.5 Partial interaction	**16**
1.5.1 Slip and slip strain	16
1.5.2 Degree of interaction	17
1.6 Buckling	**18**
1.6.1 General	18
1.6.2 Lateral-distortional buckling	19
1.6.3 Local buckling	20
1.7 References	**20**
2 Sizing of members	**21**
2.1 Introduction	**21**
2.2 Shear lag	**21**
2.2.1 General	21
2.2.2 Sizing for effective width	23
2.2.3 Effective section of a composite member	25
2.3 Local buckling	**27**
2.3.1 General	27
2.3.2 Initial local buckling	27
2.3.4 Beams in positive bending	35
2.3.5 Local buckling in shear	35
2.3.6 Concrete-filled steel tubes	36
2.4 References	**38**
3 Elastic analysis of composite beams	**39**
3.1 Introduction	**39**
3.2 Linear material properties	**39**
3.3 Full interaction analysis	**40**
3.3.1 Elastic transformed cross-sections	40
3.3.2 Continuous composite beams	44
3.3.3 Deflections due to creep	45
3.3.4 Deflections due to shrinkage	46
3.4 Partial shear connection	**47**
3.4.1 Simplified model	47
3.5 Method of construction	**48**
3.5.1 General	48
3.5.2 Flexural stresses	49
3.6 Shear flow on connectors	**51**
3.6.1 General	51
3.7 References	**52**

4 Rigid plastic analysis of simply supported beams   **53**
4.1 Introduction   **53**
4.2 Rigid plastic flexural capacity of standard composite beams   **53**
     4.2.1 Equilibrium of forces at a design section   53
     4.2.2 Steel component weakest   55
     4.2.3 Concrete component weakest   62
     4.2.4 Shear component weakest   64
     4.2.5 Effect of vertical shear on the flexural capacity   66
4.3 Rigid plastic flexural capacity of encased composite beams   **67**
     4.3.1 General   67
4.4 Variation of flexural capacity along the length of the beam   **70**
     4.4.1 General   70
     4.4.2 Uniformly distributed shear connection   70
4.5 References   **73**

5 Mechanical shear connectors   **74**
5.1 Introduction   **74**
5.2 Local detailing rules   **75**
     5.2.1 General   75
     5.2.2 Shape of stud shear connections   76
     5.2.3 Spacing of stud shear connectors   76
5.3 Dowel resistance to the shear flow forces   **77**
     5.3.1 General   77
     5.3.2 Mean strength in push-tests approach   77
     5.3.3 Characteristic strength in composite beams approach   78
     5.3.4 Composite beams with non-uniform loads   82
     5.3.5 Composite beams designed using linear-elastic theory   83
5.4 Fracture of shear connectors due to excessive slip in simply supported beams   **84**
     5.4.1 General   84
     5.4.2 Slip capacities of stud shear connectors S_{ult} 86
     5.4.3 Parametric study approach   86
     5.4.4 Mixed analysis approach   88
5.5 References   **94**

6 Transfer of longitudinal shear forces   **95**
6.1 Introduction   **95**
6.2 Shear flow planes   **95**
6.3 Shear flow forces   **95**
6.4 Generic shear flow strengths   **97**
6.5 Resistance of shear plane traversing depth of slab   **99**
     6.5.1 Shear flow strength of full depth plane   99
6.6 Resistance of shear planes that encompass connectors   **104**
     6.6.1 Strength of planes encompassing connectors   104
6.7 References   **106**

7 Stocky columns   **107**
7.1 Introduction   **107**
7.2 Plastic centroid and concentrically loaded column   **108**
7.3 General methods of analysis   **110**
     7.3.1 Elastic-plastic technique   110
     7.3.2 Rational non-linear analysis   114
7.4 Rigid plastic analysis   **115**
     7.4.1 General   115
     7.4.2 Point 'B'   116
     7.4.3 Point 'C'   118
     7.4.4 Point 'D'   119
     7.4.5 Allowance for shear   120
7.5 References   **120**

8 Slender columns   **121**
8.1 Introduction   **121**
8.2 Elastic columns   **121**
     8.2.1 Concentric loading   121
8.3 End moments   **127**

8.3.1 Secondary effects 127
8.3.2 Graphical interpretation 129
8.4 Moment capacity of slender composite columns **130**
8.4.1 Concentrically loaded columns 130
8.4.2 Second order effects 132
8.4.3 Moment capacity for a given column 132
8.5 References **134**

9 Composite beams with service ducts **135**
9.1 Introduction **135**
9.2 Outline of general analysis procedure **135**
9.3 Maximum flexural capacity of ducted beam **137**
9.4 Pure flexural capacity of ducted region **139**
9.4.1 Flexural behaviour 139
9.5 Pure shear capacity of ducted region **141**
9.5.1 Mechanism of shear transfer 141
9.5.2 Pure shear capacity of steel T-section 143
9.5.3 Pure shear capacity of composite T-section 145
9.6 Interaction between shear and flexure **150**
9.6.1 Failure envelope 150
9.7 Enhanced shear strength due to the shear resistance of the slab **151**
9.7.1 Contribution of slab 151
9.8 Strengthening ducted regions by plating **154**
9.8.1 Plating 154
9.9 Service duct near supports **156**
9.9.1 General 156
9.10 Embedment failure **159**
9.10.1 General 159
9.11 Reference **161**

10 Local splitting **162**
10.1 Introduction **162**
10.2 Mechanisms of splitting **162**
10.3 Splitting resistances of slabs with rectangular cross-sections **164**
10.3.1 Splitting resistance to individual connectors 164
10.3.2 Effective widths of prisms 165
10.4 Effective dimensions for groups of connectors **168**
10.5 Pairs of connectors **168**
10.5.1 Splitting resistance to pairs of connectors 168
10.6 Groups of connectors **170**
10.6.1 Splitting resistance to groups of connectors 170
10.7 Blocks of connectors **171**
10.7.1 Blocks of stud shear connectors 171
10.8 Prisms with non-rectangular cross-sections **172**
10.8.1 Upper and lower bound solutions 172
10.8.2 Equivalent prism concept 173
10.9 References **176**

11 Post cracking dowel strength **177**
11.1 Introduction **177**
11.2 Hooped reinforcing bars **177**
11.2.1 Dowel strength of studs with hooped reinforcement 178
11.3 Post-cracking confinement of concrete **180**
11.3.1 Dowel strength with straight transverse bars 180
11.4 Post-splitting transverse forces **182**
11.4.1 Transverse splitting forces 183
11.5 References **184**

12 Rigid plastic analysis of continuous composite beams **185**
12.1 Introduction **185**
12.2 Continuous steel beams **186**
12.2.1 The plastic hinge 186
12.2.2 Requirements for plastic analysis of steel beams 188
12.2.3 Plastic analysis of continuous steel beams 190

Contents

12.3 Continuous composite beams	**193**
12.3.1 General	193
12.3.2 Composite plastic hinges	194
12.3.3 Plastic analysis of continuous composite beams	196
12.4 References	**198**
13 Lateral-distortional buckling	**199**
13.1 Introduction	**199**
13.2 Steel componentbehaviour	**200**
13.2.1 General	200
13.2.2 Design by buckling analysis	200
13.3 Design models	**203**
13.3.1 General	203
13.3.2 Inverted U-frame approach	203
13.3.3 Empirical approach	207
13.4 Recommendations	**208**
13.5 References	**208**
14 General fatigue analysis procedures	**209**
14.1 Introduction	**209**
14.2 General fatigue properties	**209**
14.2.1 General	209
14.2.2 Fatigue endurances	209
14.2.3 Residual strength	211
14.3 Applied loads on bridges	**214**
14.3.1 General	214
14.3.2 Frequency of fatigue vehicles	215
14.3.3 Standard fatigue vehicles	215
14.3.4 Load spectrum	216
14.4 Cyclic stress resultants	**218**
14.4.1 General	218
14.4.2 Influence line diagrams	218
14.4.3 Equivalent range of cyclic forces	222
14.4.4 Force spectrum	225
14.5 Frictional shear flow resistance	**226**
14.5.1 General	226
14.5.2 Frictional resistance	227
14.5.3 Frictional resistance influence line diagrams	228
14.6 Generic fatigue equation	**231**
14.6.1 General	231
14.6.2 Generic fatigue material properties	231
14.6.3 Fatigue damage analysis	232
14.6.4 Generic fatigue equation	233
14.7 References	**234**
15 Fatigue analysis of stud shear connections	**235**
15.1 Introduction	**235**
15.2 Stud shear connector fatigue material properties	**235**
15.2.1 Crack initiation properties	235
15.2.2 Crack propagation properties	237
15.3 Details of composite beam	**237**
15.4 Crack initiation approach	**239**
15.4.1 Design mode	239
15.4.2 Assessment mode	243
15.5 Crack propagation approach	**245**
15.5.1 Design mode	245
15.5.2 Assessment mode	248
15.6 Composite building beam	**252**
15.6.1 General	252
15.7 References	**255**
Index	**256**

Preface

In a companion book entitled 'Composite Steel and Concrete Structural Members: Fundamental Behaviour[1]', the authors have described the fundamental behaviour of composite members in order to give the engineer a *feel* for the behaviour that is often missing when design is based solely on using codes of practice or by the direct application of prescribed equations. This was achieved by first describing both the basic material responses and the basic structural mechanisms, which were then used to develop the fundamental equations or fundamental analysis procedures that simulated mathematically the structural responses, and which was then consolidated with a few carefully chosen worked examples.

The aim of this book on *elementary behaviour* is to supplement the book on *fundamental behaviour[1]* in order to develop *analysis skills, familiarity* and an *intuitive feel* for composite construction that is required by students and practising engineers. A topic is first described very briefly and not comprehensively. Numerous examples are then worked and used to give an in-depth illustration of a technique, point or concept. The worked examples are therefore part of the main text, and it is necessary for the reader to work through all of them to gain a full understanding. In contrast, an engineer can obtain an in depth knowledge of the development of the techniques from the companion book on *fundamental behaviour[1]*, which also contains more advanced analysis techniques. Both books are self-contained.

This book on *elementary behaviour* describes the analysis techniques required for standard forms of composite steel and concrete structures, and in particular the analysis techniques required for non-standard forms of construction that are not or rarely covered in national standards. In fact, most of the analysis techniques described in this book are not covered by national standards. The analysis procedure is described firstly in general terms, this is then followed by detailed worked examples in which the technique is applied in the assessing and upgrading of existing structures or in the designing of new structures. The subject may therefore be of interest to practising engineers, particularly if they are involved in the design or assessment of non-standard or unusual composite structures for buildings and bridges, or are involved in the upgrading or strengthening of existing composite structures. However, this book has been written specifically for teaching elementary analysis skills to undergraduate students. Factors of safety or resistance factors are not included in the analysis procedures, and it will be left to the designer to include his or her own national values. However, mean and characteristics strengths are included as these are basic properties. It is only necessary for the student before using this book to have grasped fundamental concepts of mechanics such as equilibrium, compatibility, Young's modulus and second moment of area.

It is not the object of this book to provide quick design procedures for composite members, as these are more than adequately covered by recourse to such aids as safe load tables. The emphasis in writing this book is to develop both elementary analysis skills and a feel for composite construction through the direct design and assessment of composite structural members, and in a manner that ensures that the student or engineer understands the fundamental principles and assumptions on which the analysis procedure is based. The contents have been divided into fifteen very short self-contained chapters, many of which can be taught in single one hour lectures. By using this format, the instructor can choose chapters according to his or her interest and the length of the course.

Chapter 1 introduces in general qualitative terms different forms of composite construction, their behaviour and the terminology peculiar to this form of construction. Chapter 2 idealises the shape and size of the component by defining effective sizes that allow for shear lag, voids and local buckling. The composite structure is now ready for analysis. Standard forms of simply supported composite beams are then analysed elastically in Chapter 3, their flexural capacity is determined in Chapter 4 using rigid plastic analyses, the strength and ductility of their mechanical shear connectors is treated in Chapter 5 and the resistance of the slab of the composite beam to the mechanical shear connectors is determined in Chapter 6. Standard forms of stocky composite columns are analysed in Chapter 7 and slender composite columns in Chapter 8.

At this stage, the standard forms of analysis have now been described and applied to standard forms of composite construction. Composite beams with service ducts in the webs of the steel component, which is a common form of construction, are dealt with in Chapter 9. Longitudinal splitting of the slabs of composite beams, which is the commonest form of shear failure of the slab and which is rarely if ever dealt with in standards, is covered in Chapters 10 and 11. Chapter 12 deals with the elastic and plastic analysis of continuous composite beams as well as moment redistribution, and Chapter 13 with lateral-distortional buckling of these beams. Chapter 14 applies analysis techniques for the fatigue design of stud shear connectors for new composite bridge beams, and methods for assessing the remaining strength and endurance of stud shear connectors in existing bridge beams is covered in Chapter 15.

It is suggested that a composite course should at least include Chapters 1 to 6, as this covers the basic analysis techniques required for standard forms of composite beams. This could be followed by Chapters 7 and 8 that cover composite columns. Furthermore and if there is time, Chapter 9 on composite beams with service ducts could be included, as the analysis of this form of construction requires a thorough understanding of the first six chapters and helps to consolidate the understanding of this theory, as well as to provide an interesting practical problem as a design project.

An enormous amount of personal time has been dedicated to preparing this book at the expense of our families. The authors would like to thank their wives Suzanne and Bernie and children Robert, Allan, Nigel, Amy and Adam for their good-humoured tolerance.

Reference

Oehlers, D. J. and Bradford, M. A. (1995). Composite Steel and Concrete Structural Members: Fundamental Behaviour. Pergamon Press, Oxford.

Notation

The following notation is used in this book. Generally, only one meaning is assigned to each symbol, but in cases where more than one meaning is possible, then the correct one will be evident from the context in which it is used.

A	=	cross-sectional area; area of the free body; generic form of the fatigue damage parameter;
A_b	=	area of bottom transverse reinforcement per unit length of shear plane;
A_c	=	cross-sectional area of the concrete component;
A_{col}	=	cross-sectional area of column;
A_f	=	cross-sectional area of flange;
A_h	=	area of one arm of hooped reinforcement;
A_m	=	area of applied moment diagram in a shear span; applied moment parameter;
A_{prof}	=	cross-sectional area of profiled sheeting per unit length;
A_{rib}	=	area of individual rib of composite profiled slab;
A_r	=	area of reinforcing bars;
A_s	=	cross-sectional area of the steel component;
$(A_s)_{duct}$	=	cross-sectional area of steel component in ducted region;
A_{sh}	=	cross-sectional area of the shank of a stud connector;
A_{slab}	=	cross-sectional area of slab;
A_{sr}	=	area of longitudinal shear force diagram or longitudinal thrust in a shear span; longitudinal thrust parameter;
A_t	=	area of top transverse reinforcement per unit length of shear plane;
A_{tr}	=	total area of transverse reinforcement per unit length; $A_t + A_b$;
A_{void}	=	area of an individual void between the ribs in a profiled slab;
A_{web}	=	cross-sectional area of steel web;
a	=	constant of integration;
a_o	=	length of duct;
B	=	probability of occurrence of each weight of vehicle in a load spectrum;
b	=	breadth of plate element; width of column; width of concrete component;
b/t	=	plate slenderness;
b_a	=	effective width of patch;
$(b_a)_n$	=	effective patch width of a group of n connectors;
$(b_a)_p$	=	effective patch width of a pair of connectors;
$(b_a)_1$	=	effective patch width of an individual connector;
b_c	=	width of prism; effective width of prism;
$(b_c)_{do}$	=	effective width required to achieve triaxial restraint for dowel action;
b_h	=	width of haunch;
$(b_c)_{min}$	=	minimum effective width of prism;
$(b_c)_{min-group}$	=	minimum effective width of prism for a group of stud shear connectors;
$(b_c)_{min-one}$	=	minimum effective width of prism for an individual stud shear connector;
$(b_c)_{min-pair}$	=	minimum effective width of prism for a pair of stud shear connectors;
$(b_c)_n$	=	effective width of prism for a group of n connectors;
$(b_c)_p$	=	effective width of prism for a pair of connectors;
$(b_c)_1$	=	effective width of prism for an individual connector;
b_{eff}	=	effective breadth of slab;
b_f	=	width of concrete flange outstand; breadth of steel flange;
b_{haunch}	=	mean width of haunch;
b_i	=	effective width of pseudo inner prism;
b_l	=	breadth of slab on left side;
b_o	=	effective width of pseudo outer prism;
b_r	=	breadth of slab on right side;
b_{tr}	=	mean width of trough;
C	=	resultant force in a component; constant in the generic form of the fatigue endurance equation; constant in endurance equation that defines the mean and characteristic values;
C_w	=	strength of the weakest component in a standard composite beam;
c	=	cover;

c_b	=	bottom cover to reinforcing bar
c_{do}	=	minimum cover to stud shear connector to achieve dowel strength; $2.2d_{sh}$;
c_m	=	factor to allow for different end moments in a column;
D	=	one standard deviation; dowel strength of an individual shear connector; total depth of composite beam; depth of column;
D_{crack}	=	dowel strength of shear connector after longitudinal cracking
D_{max}	=	maximum dowel strength of an individual connector; where $(D_{max})_{push}$ is the strength in push-tests and $(D_{max})_{beam}$ is the strength in composite beams; direction of dispersal of a concentrated force; strength of shear connector prior to cyclic loading;
d	=	distance;
d_c	=	distance from top fibre to centroid of concrete component;
d_e	=	distance from top fibre to elastic centroid;
d_f	=	longitudinal spread of a group of connectors;
d_h	=	internal diameter of hoop;
d_o	=	outside diameter of circular steel tube;
d_p	=	distance from top fibre to plastic centroid;
d_r	=	diameter of reinforcing bar;
d_s	=	distance from top fibre to centroid of steel component;
ds/dx	=	slip strain;
$(ds/dx)_{duct}$	=	slip strain across duct;
$(ds/dx)_{inter}$	=	slip strain across steel-concrete interface;
d_{sh}	=	diameter of the shank of a stud shear connector;
d_{solid}	=	depth of solid portion of the slab;
d_w	=	depth of the web;
E	=	Young's modulus; endurance of a structural component;
E_c	=	Young's modulus for concrete; short term Young's modulus of concrete;
E_{ch}	=	characteristic endurance at two standard deviations;
E_e	=	effective modulus; long term modulus;
E_k	=	endurance of a component at range R_k;
E_s	=	Young's modulus for steel; usually taken as 200 kN/mm²;
E_{st}	=	initial modulus of the strain hardening range;
e	=	eccentricity of load;
EI	=	flexural rigidity;
$(EI)_{cmp}$	=	flexural rigidity of composite section;
$(EI)_e$	=	effective flexural rigidity of composite column;
$(EI)_{cmp}$	=	flexural rigidity of composite section;
$(EI)_{no}$	=	flexural rigidity of a composite beam with no interaction;
$(EI)_s$	=	flexural rigidity of steel component;
$(EI)_{slab}$	=	flexural rigidity of the slab;
F	=	force;
F_c	=	axial force in concrete component;
F_{comp}	=	compressive force;
F_{cmp}	=	transverse compressive force;
F_f	=	tensile force induced by flexure; force constant ΣfR^m;
F_n	=	normal force at interface;
F_{nf}	=	normal force to shear plane per unit length;
F_{patch}	=	concentrated load applied as a patch;
F_r	=	resultant force;
F_s	=	axial force in steel component;
F_{sc}	=	concentrated force applied by an individual shear connector;
F_{sh}	=	shear force in shear span; shear force in an individual shear connector;
F_t	=	transverse force; transverse tensile force induced by splitting;
$(F_t)_{cmp}$	=	transverse compressive force;
$(F_t)_{ten}$	=	transverse tensile force;
F_{ten}	=	tensile force; transverse tensile force;
f	=	strength; function; frequency of cyclic stress resultant;
f_b	=	bond strength of profiled sheeting ribs;
f_c	=	compressive cylinder strength of concrete; approximately $0.85f_{cu}$;
f_{cb}	=	Brazilian tensile strength; $0.5\sqrt{f_c}$ for normal density concrete;

f_{cf}	=	flexural tensile strength; $0.6\sqrt{f_c}$ for normal density concrete;
f_{ct}	=	direct tensile strength of the concrete; $0.4\sqrt{f_c}$ for normal density concrete;
f_{cu}	=	compressive cube strength of the concrete;
f_{cy}	=	compressive 'yield strength' of the concrete; $0.85f_c$ or approximately $0.72f_{cu}$;
f_{fy}	=	equivalent yield strength; flexural stress to cause yield in an element subjected to shear;
f_{max}	=	maximum transverse stress
f_u	=	ultimate tensile strength;
f_y	=	yield strength of steel;
f_{yp}	=	yield strength of profiled sheeting; proof stress;
f_{yr}	=	yield strength of reinforcing bars; maximum stress in reinforcing bar that can be achieved when not fully anchored;
f_u	=	ultimate tensile strength of the stud material;
H	=	longitudinal shear force; longitudinal compressive force; intercept of fatigue regression line; component detail parameter;
H_u	=	longitudinal compressive force at low moment end of top T-section;
H_{uh}	=	longitudinal compressive force at high moment end of top T-section;
h	=	vertical distance;
h_b	=	height of bottom steel T-section;
h_c	=	height of concrete component; height of slab; effective height of prism;
$h_{cen,c}$	=	distance between centroid of concrete component and interface;
$h_{cen,s}$	=	distance between centroid of steel component and interface;
h_{cent}	=	distance between the centroid of the concrete component and the centroid of the steel component; distance of centroid of reinforcing bar from the base of the stud;
h_H	=	lever arm between horizontal compressive forces;
h_n	=	distance of neutral axis from plastic centroid for condition of pure bending;
h_o	=	height of duct opening;
h_r	=	depth of reinforcing bars;
h_{rib}	=	height of rib of composite profiled beam; height of rib of haunch;
h_s	=	height of steel component;
h_{solid}	=	height of solid part of concrete component;
h_{st}	=	height of stud shear connector;
h_t	=	height of top composite T-section;
h_{wc}	=	height of stud weld collar;
I	=	second moment of area; second moment of area of column about weaker principal axis; second moment of area of steel beam;
I_c	=	second moment of area of concrete component;
I_f	=	second moment of area of the flange about an axis through the web;
I_{nc}	=	second moment of area of the composite section transformed to concrete taken about the centroid of the transformed concrete section; transformed second moment of area about the neutral axis;
I_{ns}	=	second moment of area of the composite section transformed to steel taken about the centroid of the transformed steel section; second moment of area of 'steel' section in negative bending;
I_s	=	second moment of area of steel component;
i	=	number of weights of fatigue vehicles; number of levels in the load spectrum;
j	=	number of fatigue zones;
K	=	shear connector stiffness or modulus; constant for determining the maximum slip in a composite beam; parameter $A_c y_c / I_{nc}$;
K_{ch}	=	parameter defining the characteristic dowel strength; $4.7 - 1.2/\sqrt{N_{gr}}$;
K_{lng}	=	long term value of parameter $A_c y_c / I_{nc}$;
K_{sht}	=	short term value of parameter $A_c y_c / I_{nc}$;
k	=	local buckling coefficient;
k_e	=	effective length factor;
L	=	longitudinal distance; span of beam; length of column; length of shear plane; length of portion of the shear span;
L_b	=	length of slab between parallel beams;
L_c	=	maximum distance between points of contraflexure;
L_{con}	=	longitudinal spacing of connectors;

L_d	=	length of beam between supports for lateral-distortional buckling;
L_{duct}	=	distance of duct from nearest support;
L_e	=	distance between edge of stud shear connector and edge of flange; effective length of a column; $L_e \geq 1.3 d_{sh}$;
L_f	=	load constant $\sum BW^m$;
L_L	=	longitudinal spacing of stud shear connectors; $0.5 d_{sh} \leq L_L \leq 6 h_c$;
L_n	=	length of shear span n;
L_p	=	area of shear plane per unit longitudinal length; perimeter length of shear plane;
L_r	=	longitudinal spacing of reinforcing bars;
L_{si}	=	longitudinal spacing of a single line of connectors; longitudinal spacing if the connectors were placed along a single line
L_{sp}	=	length of shear span; spread of reinforcing bars required to confine the concrete; spread of shear connectors that can fail as a group;
L_{ss}	=	length of shear span; length between the design section and the support in a simply supported beam;
L_T	=	transverse spacing of stud shear connectors; $L_T \geq 4 d_{sh}$;
L_{tr}	=	longitudinal spacing of transverse reinforcement;
M	=	moment; moment capacity;
M_a	=	applied moment;
$(M_a)_y$	=	applied moment to cause first yield;
$(M_a)_h$	=	applied moment at high moment end of duct;
$(M_a)_l$	=	applied moment at low moment end of duct;
M_{avl}	=	available local moment capacity;
M_{bh}	=	moment in bottom steel T-section at high moment end of duct; moment capacity of bottom steel T-section at high moment end of duct;
M_{bl}	=	moment in bottom steel T-section at low moment end of duct;
M_c	=	moment in the concrete component;
M_{cmp}	=	moment capacity of composite section when governed by distortional buckling;
$(M_{duct})_h$	=	flexural capacity at high moment end of duct;
$(M_{duct})_l$	=	flexural capacity at low moment end of duct;
M_{fsc}	=	full-shear-connection moment capacity of a composite beam;
M_{frac}	=	moment to cause fracture of the shear connection due to excessive slip;
M_{hog}	=	hogging or negative moment;
M_{int}	=	moment capacity of duct subjected to vertical shear load V_{int};
M_m	=	end moments in a column;
M_{ml}	=	maximum end moment in a column;
M_{max}	=	maximum moment; maximum value of the second order moment in a column; the sum of the primary and secondary moments;
M_o	=	pure flexural moment capacity;
M_{od}	=	the elastic lateral-distortional buckling moment in the steel component;
M_p	=	rigid plastic moment for bending about the weaker axis;
M_{ps}	=	rigid plastic moment capacity of steel component;
$(M_p)_{hog}$	=	rigid plastic moment capacity of composite beam in hogging moment;
$(M_p)_{sag}$	=	rigid plastic moment capacity of composite beam in sagging moment;
M_{psc}	=	partial-shear-connection moment capacity of a composite beam;
M_{pure}	=	pure flexural capacity;
$(M_{pure})_{duct}$	=	pure flexural capacity at mid-span of duct;
M_{rqd}	=	required moment capacity to resist shear;
M_{res}	=	reserve moment capacity;
M_s	=	moment in steel component; cross-sectional strength of the steel component in bending; moment capacity of a steel beam; rigid plastic moment capacity of steel component;
$(M_s)_p$	=	moment in steel component when composite section is fully plastic;
$(M_s)_y$	=	moment in steel component when composite section first yields;
M_{sag}	=	sagging or positive moment;
M_{sd}	=	bending strength of the steel in the absence compression;
M_{sdr}	=	reduced steel bending strength for the effects of axial compression;
M_{steel}	=	moment in steel component of a composite beam;
M_{th}	=	moment in top composite T-section at high moment end of duct;
M_{tl}	=	moment in top composite T-section at low moment end of duct;

M_{short} = short term moment taken about the plastic centroid;

M_{top} = moment taken about the top fibre;

M_y = first yield moment;

m = non-dimensional moment in a column M_{max}/M_p; slope of fatigue regression line; exponent of fatigue equation; material component parameter;

N = number of connectors in part of a span; axial force; number of cycles of load;

N_a = normal tensile force induced by V_a;

N_{col} = strength of simply supported column; strength of Euler column;

N_{crit} = elastic buckling load of a perfect simply supported column; elastic buckling load of the bottom flange in a composite beam;

$(N_{crit})_{min}$ = minimum value of elastic buckling load of the bottom flange;

N_E = Euler buckling load;

N_{FV} = number of traversals of fatigue vehicle W_{FV};

N_k = number of cycles of load of range R_k;

N_{gr} = number of connectors that can be assumed to fail as a group; in rigid plastic analysis N_{gr} can be taken as the number of connectors in a shear span N_{ss};

N_o = squash load;

N_{od} = the elastic lateral-distortional buckling load in the steel component;

N_{pure} = normal force across interface derived from the analysis of the pure shear capacity;

N_s = cross-sectional strength of the steel component in compression;

N_{short} = short term axial force;

N_{sd} = the compressive strength of the steel in the absence of bending;

N_{sq} = squash load;

N_{ss} = number of connectors in a shear span;

N_{tr} = number of connectors in a trough;

N_y = the axial load at which the column first yields;

n = modular ratio for short term loading; distance from top compressive fibre to neutral axis; depth of concrete in compression in a haunch; depth of element in compression; non-dimensional axial load in a column N_{col}/N_{sq}; number of connectors that can fail as a group; number of connectors in a group;

n_a = neutral axis position below the top fibre;

n_b = neutral axis position above the bottom fibre;

n_c = non-dimensional axial load to cause failure of concentrically loaded column;

n_e = modular ratio for long term loading;

N-A = neutral axis;

P = component strength;

P_c = strength of concrete component; $A_c 0.85f_c$; $A_c f_{cy}$; remaining strength or residual strength of a component after cyclic loading;

P_{group} = splitting resistance to a group of connectors;

$(P_{min})_{one}$ = minimum splitting resistance to an individual stud shear connector;

$(P_{min})_{pair}$ = minimum splitting resistance to a pair of stud shear connectors;

$(P_{min})_{group}$ = minimum splitting resistance to a groupof stud shear connectors;

P_{one} = splitting resistance to an individual connector;

$(P_{one})_{char}$ = characteristic splitting resistance to an individual connector;

P_{pair} = splitting resistance to a pair of connectors;

P_r = strength of reinforcing bars;

P_s = strength of steel component; $A_s f_y$; static strength of a component prior to cyclic loading;

P_{sh} = strength of shear component in a shear span; ND; $Q_{sh} L_{sp}$;

$(P_{sh})_{fsc}$ = strength of shear connection for full shear connection;

P_{split} = resistance to splitting;

$(P_{split})_i$ = resistance to splitting of inner prism;

$(P_{split})_o$ = resistance to splitting of outer prism;

P_{sq} = squash load;

p = percentage of reinforcing bars;

pf_{yr} = yield strength of the reinforcement when fully anchored per unit area of the shear plane; bond strength of the reinforcement when not fully anchored per unit area of the shear plane;

Q = shear flow strength;

Q_{ch} = characteristic shear flow strength of a shear plane;

Q_D = shear flow strength of stud shear connectors; static strength D_{max} per unit length;

Q_f = shear flow strength required to resist the fatigue loads in the crack initiation approach;

Q_{fric} = mean frictional shear flow resistance;

Q_o = shear flow strength required to resist the maximum overload;

Q_{of} = shear flow strength required to resist both the maximum overload and the reduction in strength due to fatigue damage; shear flow strength required when the structure is first built;

Q_{res} = residual or remaining shear flow strength after cyclic loading;

Q_{sh} = shear flow strength of the shear connection; strength of shear connection per unit length;

Q_{st} = shear flow strength at the start of cyclic loading; shear flow strength when first built;

q = shear flow; shear flow force; longitudinal force per unit length;

q_{do} = shear flow force resisted by the dowel action of the mechanical shear connectors;

$(q_{do})_n$ = shear flow force resisted by the dowel action of the mechanical shear connectors in shear span n;

q_o = shear flow force induced by maximum overload; maximum uni-directional shear flow force that the connector has to resist;

q_r = uni-directional shear flow force;

$(q_s)_{max}$ = maximum uni-directional shear flow force; maximum static shear flow force imposed by the traversal of the standard fatigue vehicle.

q_t = total shear flow force imposed by shear connectors; total range of the shear flow force; total range of shear flow force that causes fatigue damage;

q_v = maximum uni-directional shear flow force;

R = stress resultants; range of cyclic load; reaction;

R_{rib} = transverse rib reduction factor to the dowel strength of the stud shear connector;

R_U = nominal strength of the member;

r = radius of gyration; $\sqrt{(I/A)}$;

r_y = minor axis radius of gyration of the compressive flange;

S_{max} = maximum slip;

S_p = slip at the commencement of plasticity;

S_s = plastic section modulus; $M_{ps} = S_s f_y$;

S_{ult} = slip at fracture of the shear connector;

s = slip; longitudinal spacing of stud shear connectors;

T = transverse tensile force; total number of fatigue vehicle traversals in a design life; total number of fatigue vehicle traversals in a fatigue zone;

T_{adj} = transverse distance to adjacent beam;

T_{edge} = transverse distance to edge of slab;

TFL = fatigue zone; fatigue damage; $TF_f L_f$;

t = time; thickness of plate element;

t_f = flange thickness; thickness of plate to which the stud is welded; $t_f \geq 0.4d_{sh}$;

t_o = time at application of constant stress;

t_n = lateral spacing between connectors at the extremities of a group;

t_p = lateral spacing of a pair of connectors;

t_w = thickness of web;

u_c = longitudinal displacement of concrete component;

u_o = variation in initial imperfection in a column; additional deflection in column due to the bending curvature;

u_s = longitudinal displacement of steel component;

u_t = buckling deformation;

V = vertical shear force; magnitude of moving point load; axle load;

V_a = applied shear force;

V_b = shear force in bottom steel T-section;

V_c = shear resisted by the concrete slab;

V_{int} = shear load in combination with M_{int};

V_{mat} = material shear strength of steel web;

$(V_{mat})_b$ = material shear strength of steel web of bottom steel section;

$(V_{mat})_t$	=	material shear strength of steel web of bottom steel section;
V_n	=	shear force in shear span n;
V_{pure}	=	pure shear capacity;
$(V_{pure})_{duct}$	=	pure shear capacity of ducted section;
V_s	=	shear force in steel component;
V_t	=	shear force in top composite T-section;
$(V_t)_{upper}$	=	upper bound to pure shear capacity of top composite T-section;
v	=	deflection;
$(v_u)_{char}$	=	characteristic shear strength of shear plane;
v_{full}	=	deflection of composite beam with full interaction;
v_{no}	=	deflection of composite beam with no interaction;
v_{part}	=	deflection of composite beam with partial interaction;
W	=	W_{FV}/W_{SFV}; concentrated load;
$W_{collapse}$	=	concentrated load to cause collapse of beam;
W_{FV}	=	weight of fatigue vehicle;
W_o	=	weight of maximum overload vehicle;
W_{SFV}	=	weight of standard fatigue vehicle;
w	=	width; width of slab;
w_a	=	uniformly distributed applied load;
w_{eff}	=	effective flange width;
$(w_{eff})_{duct}$	=	effective width of slab over ducted region;
w_f	=	width of steel flange;
w_{sht}	=	short term uniformly distributed load;
w_{lng}	=	long term uniformly distributed load;
X	=	parameter in the denominator of the generic form of the cyclic stress resultant;
x	=	level of load spectrum;
x_t	=	length of lateral tensile stress distribution;
y	=	distance from top compressive fibre; vertical distance; distance from centroid of section to position of stress σ;
$\overline{y}$	=	distance between the centroid of the concrete component and the centroid of the transformed composite beam
y_c	=	depth of plastic neutral axis from the inside of the compression flange; distance between the centroid of the concrete component and the centroid of the composite section transformed to a concrete section;
y_n	=	depth of the neutral axis below the top fibre;
Z	=	structural property; elastic section modulus;
Z_{mean}	=	mean property;
Z_{ch}	=	characteristic property;
z	=	distance from end of column; number of magnitudes of the cyclic ranges; level of force spectrum; number of levels in the force constant;
α	=	exponent for the effect of span on the maximum slip; reference to a specific shear span;
α_c	=	neutral axis parameter;
α_t	=	elastic restraint stiffness per unit length applied to flange strut;
β	=	exponent for the moment effect on the maximum slip; moment gradient in a column;
γ	=	load factor; neutral axis factor;
Δ	=	change;
δ	=	additional deflection of column;
δ_o	=	maximum value of the initial out of straightness;
ε	=	strain; strain profile;
ε_c	=	strain in concrete;
ε_{cr}	=	creep strain;
ε_e	=	instantaneous strain;
ε_{fr}	=	fracture strain;
ε_s	=	strain in steel;
ε_{sh}	=	shrinkage strain;
ε^*_{sh}	=	final shrinkage strain;
ε_{st}	=	strain in steel at start of strain hardening;
ε_u	=	ultimate compressive strain of concrete; 0.003;
ε_y	=	yield strain;

η	=	degree of shear connection; strength of shear connection as a proportion of that required for full-shear-connection; imperfection parameter;
η_{max}	=	degree of shear connection at the position of maximum applied moment;
η_t	=	degree of shear connection at the transition point;
θ	=	angle of sloping side from vertical in degrees; angle in degrees between the direction of the span of the ribs and that of the composite beam; slope;
θ_{hog}	=	rotation of plastic hinge in hogging region;
κ	=	curvature;
κ_{st}	=	curvature at first strain hardening;
κ_y	=	curvature at first yield;
λ	=	slenderness ratio; L/r;
λ_d	=	buckling strength parameter for steel component in hogging bending;
μ	=	coefficient of friction at the steel-concrete interface; ≈ 0.7;
ν_s	=	Poisson's ratio for steel;
ζ	=	parameter to determine N_y;
ρ	=	density in kg/m³;
σ	=	stress; stress profile;
σ_b	=	stress in bottom fibre;
σ_{conc}	=	stress profile in concrete component;
σ_{cr}	=	maximum stress to cause elastic buckling;
σ_{equi}	=	equivalent stress profile;
σ_l	=	longitudinal stress;
σ_{max}	=	peak stress;
σ_{nc}	=	stresses in the transformed concrete section, that is the composite section transformed to concrete;
σ_{nf}	=	stress normal to shear plane; active normal stress across interface which is positive when compressive;
σ_o	=	constant stress;
σ_{real}	=	real stress profile;
τ	=	shear stress;
τ_w	=	shear stress in web; mean stress in web;
τ_y	=	shear stress to cause yield;
ϕ	=	capacity reduction factor; creep coefficient;
$\phi*$	=	final creep coefficient;
χ	=	slenderness parameter; ductility parameter;
2D	=	two dimensional dispersal of the concentrated force;
3D	=	three dimensional dispersal of the concentrated force;

1 Introduction

1.1 Composite structures

Composite steel-concrete structures are used widely in modern bridge and building construction. A composite member is formed when a steel component, such as an I-section beam, is attached to a concrete component, such as a floor slab or bridge deck. In such a composite T-beam, as shown in Figure 1.1, the comparatively high strength of the concrete in compression complements the high strength of the steel in tension. Throughout this book, we will refer to the steel and concrete as the *components* of the member, which are further made up of *elements*, such as the flanges or web of the steel I-section component, or the reinforcement in the slab.

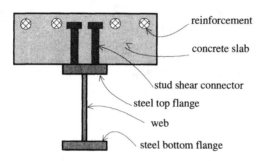

Figure 1.1 Composite T-beam

The fact that each material (steel or concrete in Figure 1.1) is used to take advantage of its best attributes makes composite steel-concrete construction very efficient and economical. However, the real attraction of composite construction is based on having an efficient *connection of the steel to the concrete*, and it is this connection that allows a transfer of forces and gives composite members their unique behaviour. In this book, we will make considerable reference to the behaviour of this connection at the interface between the steel and concrete components, and will attempt to demonstrate that the connection between the steel and concrete characterizes the composite member.

There are a number of structural arrangements in which the steel and concrete act in this symbiotic composite fashion. In simply supported bridge construction, the concrete slab is subjected to compressive forces, and this slab is supported typically by steel I-section components, as depicted in Figure 1.1. The connection between the steel and concrete is in the form of *mechanical shear connectors*, which allow the shear transfer of the forces in the concrete to the steel and vice versa, and which also prevent vertical separation of the concrete and steel components. There are many forms of mechanical shear connectors as shown in Figure 1.2. The most common, however, is the stud shear connector shown in (a), which consists of a head and a plain shank connected to the steel component by a weld collar. These stud shear connectors are considered in Chapter 5, and in bridges particularly, their efficiency may be reduced by fatigue loading, as discussed in Chapters 14 and 15.

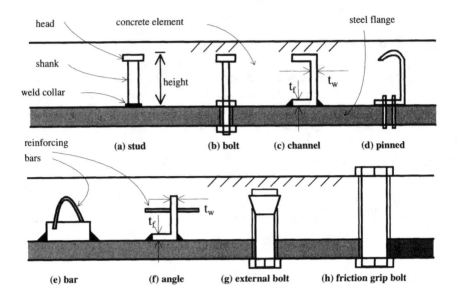

Figure 1.2   Mechanical shear connectors

It is worth noting that in composite T-beams the way in which the beam is constructed affects greatly its response to load. Buildings generally have the floors supported by closely spaced props, as shown in Figure 1.3(a), which carry all of the wet concrete loading applied to the steel component, so that the latter component does not contain any significant bending moments. This is called *propped construction*. On the other hand, environmental constraints in the construction of bridges usually prevent props from being used, as in (b), so that the flexural stiffness and strength of the steel component must carry the weight of the wet concrete. This method of construction, which is also experienced in pretensioned prestressed concrete bridge construction, is called *unpropped construction*. The ramifications of propped and unpropped construction on the flexural behaviour of beams are considered in Chapter 3.

Composite columns are also used widely in practice to resist predominantly compressive loading, and they may take the form of an encased I-section, as shown in Figure 1.4(a), a concrete-filled rectangular steel section, as in (b), or a concrete-filled steel circular tube, as in (c). The use of high strength-high performance concrete

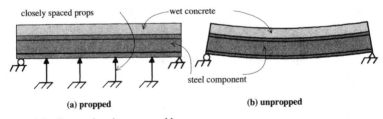

Figure 1.3   Propped and unpropped beams

is now finding its way into composite column construction, where concrete strengths may be more than twice the strength of 'normal' concrete. Short or *stocky* composite columns tend to fail by squashing, and their strength is governed by the strength of the cross-section. Stocky columns are considered in Chapter 7. Long or *slender* columns, on the other hand, tend to fail by a combination of material and geometric nonlinearities, and their strength is governed primarily by the phenomenon of buckling. Slender columns are considered in Chapter 8.

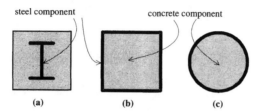

Figure 1.4 Composite column sections

Most modern flooring systems in buildings use a concrete slab with a cold formed profiled steel sheeting element about 1 mm thick as its soffit, as shown in Figure 1.5(a). This is a special form of composite member where the steel forms permanent and integral formwork for the concrete component, and the composite action is achieved by embossments in the sheeting as in (b) to (d), and by some chemical bonding between the concrete and steel sheeting. Commonly, the ribs of the profiled sheeting are orthogonal to the centreline of the I-section component which supports it, and the stud shear connectors are welded through the thin steel sheeting into the top flange element of the steel component. There is thus shear connection in the longitudinal beam direction by way of the mechanical shear connectors, as well as in the direction transverse to the steel I-beam component by the embossments in the profiled sheeting. The resulting behaviour of this system is thus referred to as *double composite action*.

Another way of forming composite beams is by filling trough girders that are fabricated from profiled sheeting with concrete as shown in Figure 1.6. The resulting

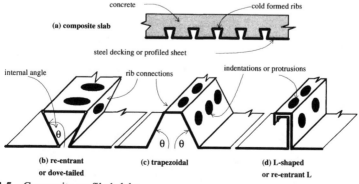

Figure 1.5 Composite profiled slabs

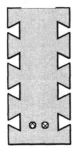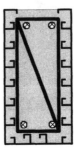

Figure 1.6 Profiled trough girders

profiled beam bears a resemblance to the composite profiled slab in that the steel sheeting is used as permanent formwork and acts compositely with the concrete. An extension of this concept in flooring construction would be to produce a flooring formwork consisting of profiled troughs and profiled sheeting, and then to pour concrete so as to produce a fully composite profiled beam-slab system.

It is now commonplace in high-rise buildings to use a combination of these composite forms of construction. For example, the columns may be concrete-filled tubes or encased I-sections, and these are connected to the core of the building by steel I-beam components. These I-beams are then made composite by laying steel profiled sheeting onto their top flange elements, welding stud shear connectors through the profiled sheeting into the flanges and pouring the concrete slab to form the flooring of the building storey. The I-sections must be connected to the columns by some form of mechanical connection, which is not to be confused with the mechanical shear connectors considered so far. Such composite connections are still the subject of vigorous on-going research, and are not treated in this book.

The composite members and composite forms of construction described in the previous discussion represent only the common applications, and the use of steel and concrete to form these types of composite member is only limited by the imagination of the designer. For example, in retrofitting deteriorating concrete beams or slabs to improve their performance, or to increase their resistance to earthquake loading, steel plates may be glued or bolted to the concrete component, and these plated members must be analysed by the theories based on composite analysis. We shall see that although this text deals with the elementary behaviour of composite members, there are a number of concepts peculiar to this form of construction, and the basic principles that are established must be borne in mind if the designer is to take advantage of composite action in his or her final design solution.

1.2 Design criteria

1.2.1 *General*

It is worth noting at this point that the main philosophy in modern structural design is based on so-called *Limit States Design* or *Load and Resistance Factor Design*. This almost universally adopted procedure was developed during the 1970's and early 1980's. Although the basis and requirements of this philosophy have been well-developed and explained in a number of texts, it will be treated briefly in Section 1.2.3. Because of the variation of material properties, particularly those of concrete, design is generally based on characteristic values, and these are introduced in Section 1.2.2. It must be remembered that in this book we are considering the *structural response* of composite members, which can loosely be defined as their behaviour when subjected to some form of applied loading. The methods of *structural analysis*, in so far as it is used to determine the loading, will not be treated specifically. This is particularly so as many of the members we will be dealing with are statically determinate, so that the distribution of the applied moments and shears can be found from statics alone. Where these internal actions need to be found by structural analysis, and this is mainly restricted to continuous beams that are statically indeterminate, the method of analysis will be described.

1.2.2 *Characteristic values*

In structural design, it is usually assumed that the frequency of the predicted properties are of a Normal type, which is sometimes termed a Gaussian Distribution. This frequency distribution is shown in Figure 1.7. If we let Z denote the predicted structural property, such as its material strength or stiffness, this Normal Distribution is characterized by (i) the mean strength Z_{mean} and (ii) its standard deviation D_Z. In order to allow for this scatter of properties, design is often simplified by basing it on the upper or lower *characteristic strength* values Z_{ch}. For properties derived from static loading, the characteristic values are defined as either the value of the property at which 5% of the values lie below, or 5% of the values lie above. This may thus be written as

$$Z_{ch} = Z_{mean} \pm 1.64 D_Z \tag{1.1}$$

where 1.64 is the number of standard deviations from the mean to reach the 5 percentile value.

The conservative value of the characteristic property is used in design. For most design procedures, such as strength and deflection calculations, this will be given by the lower characteristic value, that is $Z_{mean} - 1.64 D_Z$, as shown in Figure 1.7. Occasionally, the upper characteristic value $Z_{mean} + 1.64 D_Z$ is used, such as in crack width predictions. In limit states design it is common practice to insert the characteristic material properties into the characteristic value of the prediction equation, in order to allow for the normal scatter of both the material properties and the prediction equation. Throughout this book, it will be made clear whether the predictive equation is based on characteristic values or mean values.

It is worth noting that it is only by convention that the 5% characteristic value is used in design. In fatigue design, as in Chapters 14 and 15, the characteristic property is often defined as the value at two standard deviations from the mean, that is $Z_{ch} = Z_{mean} - 2.0 D_Z$, so that 2.3% of the values lie beyond this characteristic value.

1.2.3 *Limit states design*

We will assume implicitly in this book that design is based on limit states design principles. The basis for this method may be found in a number of books, such as Ref. 1 for concrete design and Ref. 2 for steel design.

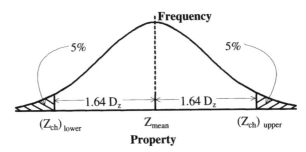

Figure 1.7   Distribution of static properties

In limit states design, the forces acting on the structure must be found, usually from loading codes. The methods of structural analysis are then used to convert these loads to the internal actions acting on the structure, and these may be the bending moments, shear forces, axial forces or even torques. These internal actions are called the *nominal actions*. For limit states design they must be multiplied by appropriate load factors γ. Because different types of loading (dead, live, wind etc.) have different variabilities and chances of occurrence, the value of the load factor for each type of loading is different. It is thus usual to multiply the appropriate load by the appropriate load factor γ at the outset, and then perform the structural analysis. The resulting (factored) action is referred to as the *design action*.

The two limit states to be considered are for *strength* and *stiffness*. For strength design, the nominal strength of the member R_U (which may be its bending, axial or shear strength) is determined, and extensive guidance is given for this calculation in this book. The nominal strength is then reduced by a capacity reduction factor f to obtain a *design strength*. The strength design equation may then be written as

$$\text{Design action} \leq \text{Design strength} \tag{1.2}$$

or

$$\Sigma \, \gamma \times (\text{Nominal action}) \leq \phi R_U \tag{1.3}$$

The load factors used in structural design vary from country to country, and it will be left to the designer to insert his or her national value. The same can be said of the capacity reduction factor. It is worth noting here that specifying the capacity reduction factor φ for composite construction is not as straightforward as in steel or concrete design alone, as we are using two different materials that have different properties. In this book, as mentioned earlier, only the strength of the member R_U will usually be specified.

The serviceability limit state is generally governed by limiting excessive deflections and vibrations. Generally for the serviceability limit state, the nominal loads are used and so are not factored, although sometimes a load factor g less than unity may be specified for long-term loads. It will be reiterated finally that this book is not a text on limit state design of composite structures, and considers only the behaviour or response of a composite member to loading. If the book is to be used in a design mode, then the reader will have to convert the governing equations to his or her national limit states format.

1.3 Material properties

1.3.1 *General*

Composite members normally use structural grade steel and normal strength concrete. However, high strength concrete may be used particularly in columns, as well as cold-rolled profiled steel sheeting in profiled slabs and beams. Detailed information may be obtained on the behaviour of these materials in a number of standard texts that deal with concrete[1] and steel[2] or Ref. 3 which treat both static and fatigue behaviour. However, a very brief description of the mechanical properties of both hot and cold-rolled steel, reinforcing steel and normal strength concrete is given in this section for use throughout this book. Stud shear connectors are treated in Chapter 5.

1.3.2 *Structural steel*

Structural steel is hot-rolled, and may be further rolled into structural shapes, most typically I-sections, or welded from flat plate to form structural sections. A typical stress-strain curve for mild steel is shown in Figure 1.8. The response is elastic-plastic-strain hardening, and the most important characteristics of the steel are its elastic modulus $E_s = 200$ kN/mm² and its yield stress f_y, typically in the range 250 N/mm² to 400 N/mm². Strain hardening generally takes place at a strain of 10 or 11 times the yield strain $\varepsilon_y = f_y/E_s$, and the initial modulus of the strain hardening range is often taken as $E_{st} = E_s/30$. The stress-strain curve is normally the same in tension and compression, and the Poisson's ratio in the elastic range is $v_s = 0.3$. In addition, the ultimate tensile stress f_u in Figure 1.8 is usually between 400 N/mm² and 500 N/mm².

The stress-strain curve shown in Figure 1.8 is based on uniaxial loading of a mild steel specimen. Sometimes the steel may be loaded in a biaxial stress state with shear stresses, and recourse is usually made to the von Mises yield criterion[2] to define the interaction between these stresses at failure. When such a yield criterion is adopted, the yield stress of mild steel in pure shear is $\tau_y = f_y/\sqrt{3}$.

1.3.3 *Profiled steel*

Profiled steel sheeting used in composite profiled slabs and beams is manufactured by cold rolling thin steel plate into an appropriate shape. The cold-rolling process tends to increase the yield stress. The stress-strain curve is rounded in the vicinity of

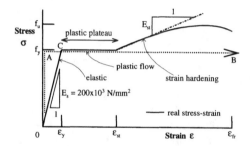

Figure 1.8   Stress-strain curve for mild steel (not to scale)

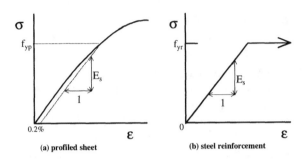

Figure 1.9 Stress-strain curves for profiled sheeting and reinforcement

the yield stress, as in Figure 1.9(a), and so a 0.2% proof stress $f_{yp} = 550$ N/mm² is often used. The elastic modulus is again close to $E_s = 200$ kN/mm².

1.3.4 *Reinforcing steel*
The concrete component is almost invariably reinforced with steel in some way. The reinforcement is usually in the form of deformed bars, or smaller diameter rectangular mesh or fabric. Here the stress-strain curve is usually elastic-perfectly plastic, as in Figure 1.9(b). Again, the elastic modulus $E_s = 200$ kN/mm², while a common value for the yield stress is $f_{yr} = 400$ N/mm².

1.3.5 *Concrete*
1.3.5.1 Short-term properties
The property of concrete that is most quoted is its compressive strength f_c. Because of the high variation of concrete strengths, it is common to specify a characteristic strength f_c' as in Section 1.2.2 that is exceeded by 95% of samples tested in compression, and so is 1.64 standard deviations below the mean strength. It will be left for the reader of this book to insert his or her characteristic values into the design equations. The concrete compressive strength f_{cu} used in these equations will be the cylinder strength, as opposed to the cube strength f_{cu} used in some countries, however, it can be assumed that $f_c \approx 0.85 f_{cu}$. The values of f_c for normal strength concretes are in the range 20 N/mm² to 40 N/mm². High strength concretes are gaining popularity in column construction, and strengths may exceed $f_c = 100$ N/mm².

The tensile strength of concrete is very much lower than its compressive strength. The direct tensile strength f_{ct} is obtained from simple pull tests and can be obtained from the empirical equation $f_{ct} = 0.4\sqrt{f_c}$ where the units are in N/mm². Lateral tensile stresses are encountered in beams with stud shear connectors, and hence we will need to know the splitting strength f_{cb}. This can be obtained directly from a Brazil strength test or approximately from $f_{cb} = 0.5\sqrt{f_c}$. The modulus of rupture f_{cf} is the flexural strength of an unreinforced concrete prism tested in flexure, and is given by the empirical expression $f_{cf} = 0.6\sqrt{f_c}$.

The stress-strain relationship for concrete under compression is very different in shape from that of steel, and depends greatly on the compressive strength f_c and also significantly on the rate of straining. A typical curve is given in Figure 1.10. At stresses below $0.4f_c$, the behaviour is close to linear elastic, with an elastic modulus E_c often quoted as

$$E_C = 0.043\rho^{1.5}\sqrt{f_C} \tag{1.4}$$

where f_c and E_c are expressed in N/mm² and ρ is the density in kg/m³. For normal weight concretes, $\rho = 2400$ kg/m³, and so $E_c = 5050\sqrt{f_c}$.

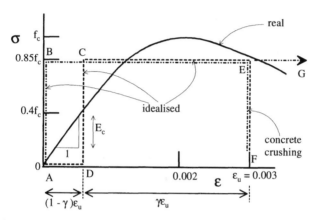

Figure 1.10   Stress-strain curve for normal strength concrete

The maximum stress f_c occurs at a strain of about 0.002 after which the stress reduces. For unrestrained concrete, it is usual to assume that the concrete fails in crushing at a strain $\varepsilon_u = 0.003$.

1.3.5.2   Long-term properties
Unlike steel, concrete is subjected to time-varying deformations due to creep, shrinkage and thermal strains. At a constant temperature, the total strain at any time t may be taken as

$$\varepsilon(t) = \varepsilon_e(t) + \varepsilon_{cr}(t) + \varepsilon_{sh}(t) \tag{1.5}$$

where ε_e is the instantaneous strain, ε_{cr} is the creep strain and ε_{sh} is the shrinkage strain. The deformations which take place under a constant stress applied at a time t = t_o are shown in Figure 1.11.

The reader is referred to specialist texts[4] on details of the mechanisms of creep and shrinkage. We will simply note here that creep is the deformation that occurs

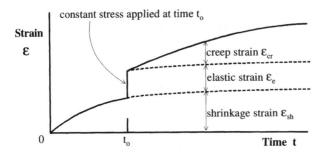

Figure 1.11   Time-dependent deformations under sustained stress

under a sustained stress applied at time t_o. It is often convenient to express the creep strain ε_{cr} in terms of the creep coefficient ϕ by

$$\phi(t,t_o) = \frac{\varepsilon_{cr}(t,t_o)}{\varepsilon_e(t)} \tag{1.6}$$

The final creep coefficient $\phi^*(t_o)$ as $t \to \infty$ is usually in the range 1.5 to 4.0.

Under the action of a constant stress σ_o, the sum of the elastic and creep strains is $\sigma_o/E_e(t,t_o)$, where E_e is the effective modulus given by

$$E_e = \frac{E_c}{1+\phi} \tag{1.7}$$

Shrinkage is also treated in standard texts[4] on time effects in concrete, and the rate of shrinkage decreases as time increases, as shown in Figure 1.12. The final shrinkage strain ε^*_{sh} as $t \to \infty$ may be as high as 1000×10^{-6}. The coefficient of thermal expansion for concrete is also of importance, and a value of $10\times10^{-6}/°C$ is often quoted for design.

1.3.6 *Rigid plastic properties*

When conducting a service load analysis (to satisfy the serviceability limit state) as is undertaken in Chapter 3, the steel and concrete components are loaded in the

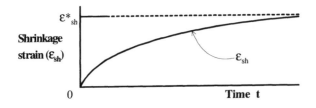

Figure 1.12   Shrinkage strains

linear elastic range and the behaviour is governed primarily by the values of the elastic moduli E_s and E_c. Hence the stress-strain curves shown in Figures 1.8 to 1.10 are invoked in their elastic ranges of structural response.

Elastic or elastic-plastic analyses of composite members can be quite tedious, and the behaviour of a composite member at ultimate may be more easily determined by conducting a *rigid plastic analysis* which assumes that the composite beam is *ductile*. A rigid plastic analysis allows the designer to calculate an upper bound to the ultimate strength, and is a prerequisite for the analysis treated in detail in Chapter 4 and used in Chapter 12. In order to conduct a rigid plastic analysis, the stress-strain relationships in Sections 1.3.2 to 1.3.5 must be simplified. This requires the assumption to be made that the materials are either not stressed at all, or are fully yielded with an infinite deformation capacity at the yield stress or plastic plateau, as shown in Figure 1.13.

The rigid plastic simplification of the steel stress-strain curve is shown as the dotted line 0-A-B in Figure 1.8. This is applicable to a composite beam whose steel component is subjected to large curvatures, so that the strains (both tensile and compressive) over most of the steel component exceed ε_y. As the fracture strain ε_{fr} in Figure 1.8 can be orders of magnitude greater than the first yield strain ε_y owing to the ductility of the mild steel, it is very unlikely that the steel will fracture before the concrete crushes, and so the assumption of an infinite plastic plateau as depicted in Figure 1.13 is adequate. Although it may seem slightly unconservative to treat the small elastically strained portions of the steel component as being fully yielded at f_y, this is compensated for by the increase in strength due to strain hardening which is ignored in the rigid plastic assumption. Hence the rigid plastic strength which is theoretically unattainable, as it requires an infinite curvature, can be attained in practice due to strain hardening.

Similarly, the rigid plastic assumption for the concrete stress-strain curve is shown as the dotted line A-B-E in Figure 1.10. Here the tensile strength of the concrete is assumed to be zero, and all compressive concrete is fully yielded with an unlimited plastic plateau at $0.85f_c$, as shown in Figure 1.13. This is in contrast to the analysis of reinforced concrete beams[1] that make use of a so-called neutral axis γ factor approach in which the concrete is not fully yielded over its entire compressive region at ultimate as shown by the line A-D-C-E-F in Figure 1.10. It is argued in Ref. 3 that use of a γ

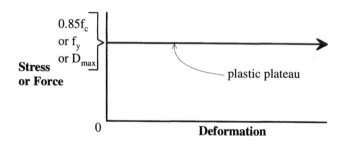

Figure 1.13 Idealized rigid plastic material properties

factor is not necessary in composite construction. It is worth noting that the 0.85 factor for the fully yielded strength of the concrete can be attributed to a size effect[1], and arises because a typical concrete component crushes at 0.85 times its cylinder strength f_c. Although details of the material properties of stud shear connectors are deferred until Chapter 5, in a rigid plastic analysis they too are assumed to deform in a rigid plastic manner with a yield plateau equal to their maximum dowel strength D_{max} as shown in Figure 1.13.

1.4   Partial shear connection

1.4.1 *General*

Two commonly-used terms that describe composite behaviour are *partial-shear-connection* and *partial-interaction*, and these relate to the behaviour of the connection between the steel and concrete components. We shall see that partial-shear-connection concerns equilibrium of the forces within a composite member, while partial-interaction concerns compatibility of deformations at the steel/concrete interface. Partial-shear-connection thus represents a *strength* criterion, while partial interaction represents a *stiffness* criterion. Numerical examples that illustrate partial-shear-connection are given in Chapter 4.

1.4.2   *Equilibrium of forces*
1.4.2.1   Composite beam

Consider the simply supported composite T-beam shown in Figure 1.14(a) and (b) that is subjected to positive or sagging bending, and we will assume that the curvature is large so that the strains are large. If we know the distribution of strains across the section A-A distant L_{sp} from the support, the stress-strain curves for the materials could be invoked to determine the stress distribution, as shown in (d). Let us assume that the neutral axis N-A lies in the steel component, so that the portion below N-A in (d) in the steel is subjected to tension and that above it is subjected to compression, and all of the concrete component is subjected to compression. If we integrate the tensile stress distribution, then this will be equivalent to a tensile force F_{ten} positioned h_2 below the steel/concrete interface, and integrating the compressive stress produces a compressive force positioned h_1 above the interface. This is shown in Figure 1.14(e). Clearly from horizontal equilibrium, $F_{comp} = F_{ten} = F_r$ and from rotational equilibrium $M_a = F_r(h_1 + h_2)$, where M_a is the applied moment at section A-A.

Another way of visualizing the internal stress resultants is to replace the stress distribution acting on the concrete component by a moment M_c and an axial force F_c acting at the centroid of this component, and to replace the stress distribution acting on the steel component by a moment M_s and an axial force F_s acting at the centroid of the steel component, as in Figure 1.14(f). Clearly then from force and moment equilibrium

$$F_c = F_s \tag{1.8}$$

$$M_a = M_c + M_s + F_c(h_{cen,c} + h_{cen,s}) \tag{1.9}$$

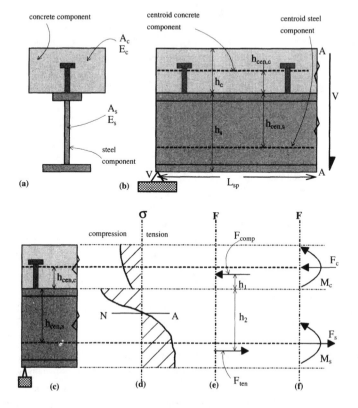

Figure 1.14 Internal forces in a composite beam

where $(h_{cen,c} + h_{cen,s})$ is the distance between the component centroids as shown in (c). In many ultimate strength design calculations (Chapter 4) it is assumed that the concrete is unreinforced in the longitudinal direction so that its strength is governed by its very low strength in tension (see Section 1.3.5). In this case it is often assumed that $M_c = 0$.

1.4.2.2 Concrete component
Consider the free body diagram of the concrete component loaded externally over the span length L_{sp} shown in Figure 1.15. As the left hand end of the span is at a simple support (or point of contraflexure), the total shear force on the shear connectors $F_{sh} = F_c$. In addition, the couple at the right hand side formed from M_c and $h_{cen,c}F_c$ must be equilibrated by the couple L_1F_n shown. The shear connectors must therefore be designed to resist the tensile normal force F_n, while the compressive normal force F_n is resisted by the shear connectors and bearing at the interface.

1.4.2.3 Degree of shear connection
So far we have considered the actions that act on a composite cross-section. We will now concentrate on the *strengths* of the components of a composite section by using

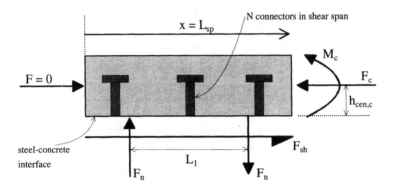

Figure 1.15 Concrete component

the rigid plastic assumptions introduced in Section 1.3.6. Firstly, the axial strength of the steel component of area A_s is defined as $P_s = A_s f_y$, where f_y is the yield stress of the steel (Section 1.3.2). The axial strength of the concrete component of area A_c is taken as $P_c = 0.85 A_c f_c$, where f_c is the compressive strength of the concrete (Section 1.3.5). Finally, we will define the strength of the shear connection, P_{sh}, as the product of the number N of shear connectors in the shear span as shown in Figure 1.15 and the strength of an individual connector D_{max}.

There are three possible stress distributions that can occur for a composite beam at its maximum strength, and these are shown in Figure 1.16 for the beam shown in Figure 1.14 under the assumption that $M_c = 0$. In Case 1 in Figure 1.16, we are assuming that $P_s < P_c$ so that the steel component is fully stressed and the concrete component is partially stressed as shown in (b). Hence $F_s = F_c = P_s$ as shown in (a), and as we saw in Figure 1.15 that $F_{sh} = F_c$, the strength of the shear connectors to ensure that this equilibrium condition exists is $P_{sh} > P_s$. This condition is referred to as one of *full-shear-connection*. The moment capacity of the section is then

$$M_{fsc} = P_s h_1 \tag{1.10}$$

In Case 2 of Figure 1.16, $P_c < P_s$ so that the concrete component is now fully stressed and therefore $F_s = F_c = P_c$. In order for this latter condition to be realized, some of the steel must be in tension and some in compression as shown in (e), resulting in the couple M_s shown in (d). Again, with the necessary equilibrium condition that $P_{sh} > P_c$, we have a situation with *full-shear-connection* and now

$$M_{fsc} = M_s + P_c h_3 \tag{1.11}$$

Note that in both Cases 1 and 2 in Figure 1.16, there is only one neutral axis.

Consider now the case where the strength of the shear connection governs, that is $P_{sh} < P_c$ and $P_{sh} < P_s$ as shown in Case 3 of Figure 1.16. For the equilibrium condition in Figure 1.15 to exist, then $F_c = P_{sh}$ so that not all of the concrete in Figure 1.16(h)

Case 1: $P_s < P_c$; $P_{sh} > P_s$; full shear connection

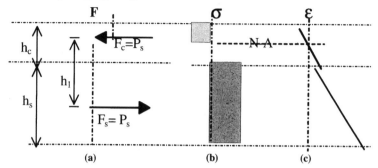

(a)   (b)   (c)

Case 2: $P_c < P_s$; $P_{sh} > P_c$; full shear connection

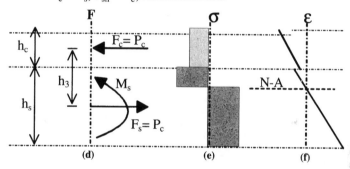

(d)   (e)   (f)

Case 3: $P_{sh} < P_c$; $P_{sh} < P_s$; partial shear connection

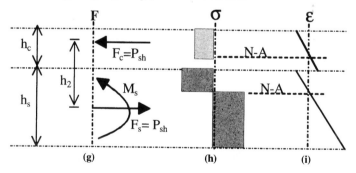

(g)   (h)   (i)

Figure 1.16   Degree of shear connection

is at its maximum stress and a neutral axis will lie in the concrete component. Furthermore as $F_s = P_{sh}$ in (g) then some of the steel is in tension and some in compression as shown in (h), and another neutral axis will lie in the steel component and a moment M_s will be induced. This is referred to as *partial-shear-connection* because the flexural capacity is now governed by the lack of shear connection, whereas in the previous cases it was assumed that there was a plentiful supply of shear connectors. The bending strength is

$$M_{psc} = M_s + P_{sh}h_2 \tag{1.12}$$

It can thus be seen in Figs. 1.16(c), (f) and (i) that there is always one neutral axis when there is full-shear-connection, and there are always two neutral axes when there is partial-shear-connection.

In Case 1 in Figure 1.16, we required that for full shear connection $(P_{sh})_{fsc} = P_s$ and similarly in Case 2 for full shear connection that $(P_{sh})_{fsc} = P_c$. On the other hand, the strength of the shear connection P_{sh} controlled the strength of the composite beam for partial interaction (Case 3). In this book, we will use the *degree of shear connection* η in a shear span, defined as

$$\eta = \frac{P_{sh}}{\left(P_{sh}\right)_{fsc}} \tag{1.13}$$

1.5 Partial interaction

1.5.1 Slip and slip strain

The behaviour of a composite beam is affected directly by the slip of the shear connection at the steel/concrete interface. The elevation of a simply supported composite beam is shown in Figure 1.17(a). When the composite beam is unloaded, the sections AB in the concrete component and CD in the steel component are in line, and positioned at some distance L from a convenient reference axis. On application of the load F, the section deforms as shown in (b). The flexural forces in the top fibres of the concrete component and steel component cause these fibres to contract, while the flexural forces in the bottom fibres of the concrete and steel cause these fibres to expand. There is thus sliding action at the interface, and the relative movement at the interface caused by this sliding action is referred to as the *slip* s.

If the new position of B in the concrete component is at $L + u_c$, as shown in Figure 1.17(b), and that of C in the steel component is at $L + u_s$, then $s = u_c - u_s$. This slip is resisted by the longitudinal shear forces. If we now consider the distribution of strains in the concrete and steel components over the length L, as in Figure 1.17(c), then

$$u_c = \int_L \varepsilon_c dx \qquad \text{and} \qquad u_s = \int_L \varepsilon_s dx \tag{1.14a,b}$$

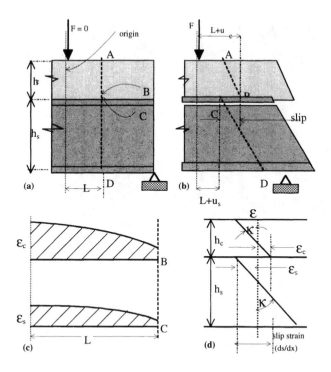

Figure 1.17   Slip and slip strain

and substituting these values into $s = u_c - u_s$ and upon differentiation,

$$\frac{ds}{dx} = \varepsilon_c - \varepsilon_s \tag{1.15}$$

The derivative of the slip ds/dx is referred to as the *slip strain* and as can be seen in Figure 1.17(d) it is the step change between the strain profiles in each component.

1.5.2 *Degree of interaction*

A condition of *no interaction* is achieved when the interface is greased, but when the steel and concrete components are in contact and so have the same curvature, as shown in Figure 1.18(a). On the other hand, when the interface is glued then $\varepsilon_c = \varepsilon_s$ and so the slip strain $ds/dx = 0$ as in (b). This condition is referred to as one of *full interaction*, and clearly *partial interaction* is the usual condition encountered between full interaction and no interaction as shown in (c).

It should be noted that the degree of interaction is a stiffness-based property, and is *not* the same as the degree of shear connection considered in Section 1.4 that is based on strength. The degree of shear connection and degree of interaction are directly related, however, as increasing the number of shear connectors both increases

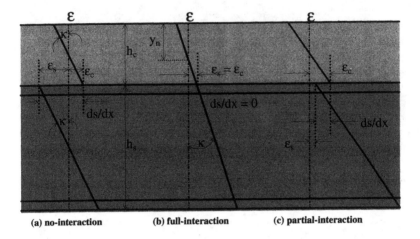

(a) no-interaction   **(b) full-interaction**   **(c) partial-interaction**

Figure 1.18   Degree of interaction

the shear strength at the interface P_{sh} and increases the shear stiffness at the shear connection. Note also that slip strains in beams with partial shear connection as shown in Figure 1.16(i) tend to be significantly larger than those in beams with full shear connection as shown in Figures 1.16(c) and (f).

1.6 Buckling

1.6.1 *General*

Although in a composite member the best use of the steel is made when it carries tensile forces, there are some cases where some of the steel is subjected to compression. For example: T-beams in negative bending (such as over an interior support or adjacent to a column) have their bottom flange element and substantial portions of the web element in compression; beams with full shear connection where the strength of the concrete element P_c governs as in Case 2 in Figure 1.16; and beams with partial shear connection have the top flange element subjected to compressive actions as shown in Case 3 in Figure 1.16.

The disadvantage of a steel element subjected to compression is that it is prone to *buckle*. The buckling of steel structures is covered in depth in standard texts[2], and essentially arises because the steel component attains a more favourable equilibrium position when it buckles or moves out of the plane of loading. In composite members, the two modes of buckling encountered are known as *local* and *lateral-distortional*, and these are covered in the following sub-sections. Buckling of the steel component usually exhausts its strength and results in catastrophic failure of a composite member, and therefore means must be established to ensure that buckling does not occur. Of course, buckling must not occur if a composite beam is analysed by using rigid plastic assumptions.

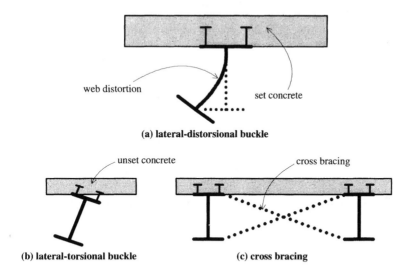

(a) lateral-distorsional buckle

(b) lateral-torsional buckle

(c) cross bracing

Figure 1.19 Lateral buckling

1.6.2 *Lateral-distortional buckling*

When a composite T-beam is subjected to negative or hogging bending, the bottom flange element is loaded in compression, and is restrained only by the stiffness of the steel web. In this lateral-distortional buckling mode, shown in Figure 1.19(a), the flange element buckles sideways and twists, with the web element distorting in the plane of its cross-section. Generally the flange element is quite stocky, so that it displaces and twists as a rigid body during buckling, with only the web element experiencing distortion during the buckling phenomenon. Distortion of the web element occurs necessarily because the top of the web is attached to the concrete component by the shear connection, and the high stiffness of the concrete component permits only very small twists during buckling. Lateral-distortional buckling depends on the moment M_s, shear force V_s and axial compression F_s that are present in the steel component. Its accurate prediction is quite complex, and recourse usually has to be made to a finite element computer program in lieu of approximate techniques.

Lateral-distortional buckling is treated in more detail in Chapter 13, where an approximate method of prediction is introduced. If we are to take advantage of rigid plastic design for continuous beams, as in Chapter 12, then lateral-distortional buckling must be prevented from occurring before the ultimate load is reached. This is usually achieved by the provision of cross-bracing, as in Figure 1.19(c). It is worth noting that lateral-torsional buckling can occur in positive bending prior to the concrete setting as shown in (b).

1.6.3 *Local buckling*

Local buckling occurs when the steel component forms 'ripples' with a short half-wavelength over the portion of the steel component in compression. In a composite T-beam in hogging bending, it may occur in the flange element and compressive portion of the web element prior to lateral-distortional buckling, as shown in Figure 1.20. Local buckling may also occur when the steel is in contact with the concrete, such as in the flange element of a T-beam in positive bending when the flange is subjected to compression, or in the thin profiled sheeting that is used to make a composite profiled slab, as in Figure 1.21. Unlike lateral-distortional buckling, local buckling in some cases does not usually cause immediate catastrophic failure, and there is often a postbuckling reserve of strength before ultimate conditions are reached.

Generally, local buckling can be prevented by imposing geometrical constraints on the steel component, such as limiting the width to thickness ratio of the flange element or the depth to thickness ratio of the web element. These constraints are used to size the member, and are discussed in the following chapter.

1.7 References

1. Warner, R.F., Rangan, B.V., Hall, A.S. and Faulkes, K.A. (1998). *Concrete Structures*. Longman, Melbourne.
2. Trahair, N.S. and Bradford, M.A. (1998). *The Behaviour and Design of Steel Structures to AS4100*. 3rd edn, E&FN Spon, London.
3. Oehlers, D.J. and Bradford, M.A. (1995). *Composite Steel and Concrete Structural Members: Fundamental Behaviour*. Pergamon Press, Oxford.
4. Gilbert, R.I. (1988). *Time Effects in Concrete Structures*. Elsevier, Amsterdam.

2 Sizing of members

2.1 Introduction

The analysis of a composite member is often not based on the gross cross-sectional proportions. In reinforced concrete design, it is common to consider only the *effective width* of a T-beam, while in slender steel plate structures the effective width is again commonly used. In composite construction, both steel and concrete are used, and so effective widths are often specified for the concrete component as well as for the steel component. The effective width treatment of the concrete component arises primarily from the effects of shear lag, while that in the steel component arises mainly from the effects of local buckling. Both of these phenomena are nonlinear, and simplifications are fortunately available for transforming the nonlinearities into a form suitable for a linear analysis. This transformation is possible by considering the *effective size* of a composite member, obtained from the effective widths of the concrete and steel components.

The methods presented in this chapter are simplifications by which the effective size (or effective section) may be determined. Once this has been determined, the section may be analysed by the methods presented in the remainder of this book. Of course, the effective size is only an analytical approximation for obtaining section properties, and it must be remembered in calculating design actions that the full load may act over the gross section, and not just the effective section.

2.2 Shear lag

2.2.1 *General*

The conventional or engineering theory of bending assumes that plane sections remain plane, which means that shearing strains are neglected. The term *shear lag*[1] is used to describe the discrepancies between the approximate engineering theory, and the real behaviour that results in both increases in the stresses in the concrete component adjacent to the steel I-section component in a composite T-beam, and to decreases in the stresses in the concrete component away from the steel.

Consider the simply supported T-beam with a central concentrated load shown in Figure 2.1(a). The shear flow distribution in the slab is linear, and this produces *warping* displacements or complementary displacements in the longitudinal direction that are parabolic in the transverse direction. In the left hand side of the beam, the shear is positive and the warping displacements are as shown in (b). On the other hand, the right hand side of the beam is subjected to negative shear, resulting in the warping displacements also shown in (b). In order for geometric compatibility to be maintained at midspan, changes are required in the bending stress distribution as well as in the shear stress distribution. These changes in stress result in the shear lag effect.

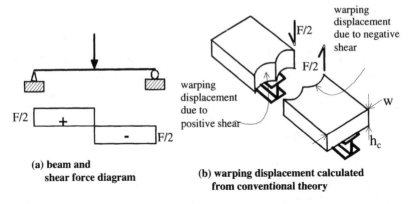

Figure 2.1 Incompatible warping displacement at a shear discontinuity

The approximate method of accounting for shear lag is to use an effective width concept, whereby in theory the actual width w of the slab is replaced by a reduced width w_{eff} given by

$$\frac{w_{eff}}{w} = \frac{\text{nominal bending stress}}{\text{maximum bending stress}} \qquad (2.1)$$

This approach simply replaces the actual bending stresses by constant stresses that are equal to the actual maximum stress distributed over an effective flange width w_{eff}.

Equation 2.1 may be restated in terms of the peak stress σ_{max} and the longitudinal stress σ_l that varies with x along the width of the concrete component. In order to allow for a nonuniform distribution of stress due to shear lag, we assume that the concrete component is narrower so that the rectangular stress block of area $w_{eff} \times h_c \times \sigma_{max}$ is equal to the area under the parabolic stress block σ_l over the width w. This is equivalent to integrating the rigorously calculated longitudinal stress in the concrete slab over the width w, and dividing by the peak value of the stress σ_{max}.

Mathematically, this restatement of Eq. 2.1 can be written as

$$w_{eff} = \frac{\int_{b_\ell}^{b_r} \sigma_\ell dx}{\sigma_{max}} \qquad (2.2)$$

where the breadths b_l and b_r are half of the transverse spans of the slab on the left and right of the steel component, as shown in Figure 2.2, and x is the coordinate transverse to the centreline of the steel component.

The shear lag problem is complex, and a particular model for the effective width may be accurate for predicting deflections yet be quite inaccurate for predicting

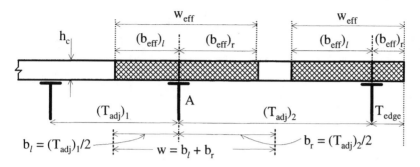

Figure 2.2 Effective width

flexural stresses. As with most of structural design, the effective width concept is only justified if the design is conservative, so that the stresses and deflections derived from linear elastic analysis using the effective width are greater than the values calculated rigorously. Fortunately, rigid plastic analyses are not overly sensitive to errors in the effective width.

2.2.2   *Sizing for effective width*
2.2.2.1 General
There are a number of parameters that affect the effective width of the concrete component of a composite beam, and as noted earlier an effective width model that is accurate for deflections may not have the same accuracy for determining flexural stresses. Because of these variations, the simplified model of the Eurocode 4[2] and Ansourian's approach[3] will be treated here. The Eurocode recommendation is that w_{eff} be calculated from

$$w_{eff} = 0.25 L_c \tag{2.3}$$

where L_c is defined as the maximum distance between points of contraflexure, and recommended values are given in Figure 2.3. Of course, the geometrical constraints

$$w_{eff} \leq \frac{\left(T_{adj}\right)_1 + \left(T_{adj}\right)_2}{2} \tag{2.4}$$

and $w_{eff} \leq 2 T_{edge}$ (2.5)

must apply, where T_{adj} and T_{edge} are shown in Figure 2.2.

The recommendation of Ansourian[3] is slightly more complex, being based on sophisticated numerical modelling. For a continuous beam, this proposal is

$$\frac{b_{eff}}{T_{adj}/2} = 1.0 - 1.2 \left(\frac{T_{adj}}{L_c}\right) \quad \text{when} \quad \frac{T_{adj}}{L_c} \leq 0.5 \tag{2.6}$$

$$b_{eff} = 0.1 L_c \qquad \text{when} \qquad \frac{T_{adj}}{L_c} > 0.5 \qquad (2.7)$$

where b_{eff} and T_{adj} refer to the value at the same side of the steel component A in Figure 2.2 and where $w_{eff} = (b_{eff})_l + (b_{eff})_r$. For a simply supported steel beam, Ansourian's recommendation is that

$$\frac{b_{eff}}{T_{adj}/2} = 1.0 - 0.6\left(\frac{T_{adj}}{L_c}\right) \qquad \text{when} \qquad \frac{T_{adj}}{L_c} \leq 1.0 \qquad (2.8)$$

$$b_{eff} = 0.2 L_c \qquad \text{when} \qquad \frac{T_{adj}}{L_c} > 1.0 \qquad (2.9)$$

where geometrical constraints similar to Eqs. 2.4 and 2.5 of course apply.

2.2.2.2 **Example 2.1** *Effective widths of slab to the Eurocode recommendations*

The composite T-beam shown in Figure 2.2 has $(T_{adj})_1 = (T_{adj})_2 = 2000$ mm and is continuous between points of contraflexure with $L_c = 7$ m. Hence from Eq. 2.3, $w_{eff} = 0.25 \times 7000 = 1750$ mm. The effective width each side of the steel component is thus $1750/2 = 875$ mm $< 2000/2 = 1000$ mm. The slab of the T-beam is therefore not fully effective, and for analysis the regions 875 mm each side of the centreline should be

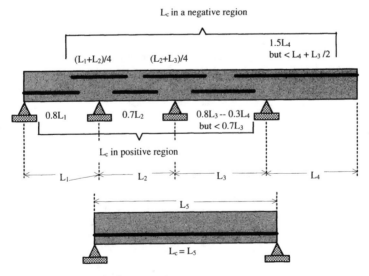

Figure 2.3 Eurocode 4 approach

considered for analysis. Of course, the dead load due to the entire slab, and any live loads over it, should be considered in determining the actual loading on the beam.

2.2.2.3 **Example 2.2** *Effective widths of slab to Ansourian's recommendations*

The same cross-section analysed in Example 2.1 according to the Eurocode is now considered using Ansourian's recommendations. Since $T_{adj}/L_c = 2000/7000 = 0.286 < 0.5$, Eq. 2.6 applies and $b_{eff}/(2000/2) = 1.0 - 1.2 \times (2000/7000) = 0.657$ so that $(b_{eff})_l = (b_{eff})_r = 657$ mm. This is more conservative than the value of 875 mm calculated in Example 2.1 using the Eurocode approach, but it is shown in Section 4.2.2.2 that ultimate strengths are insensitive to this discrepancy in effective width recommendations.

2.2.3　*Effective section of a composite member*

The concrete component in a composite member may have a profiled soffit, so that when the steel component acts compositely with such a concrete component, then the cross-sectional shape of the composite slab, to be used in the analysis, depends on the relative direction of the span of the ribs of the concrete slab to the span of the steel component.

　　The cross-section of a composite member in which the profile ribs span in the same direction as the composite beam is shown Figure 2.4(a), where h_{solid} is the height of the solid part of the concrete component and h_{rib} is the height of the rib. Also in this figure, A_{rib} is the area of an individual rib, A_{void} is the area of an individual void between the ribs as shown, and $\theta = 0°$ where θ is the angle in degrees between the direction of the span of the ribs and that of the composite beam, as shown in Figure 2.5. The cross-section can be analysed as shown in Figure 2.4(b), where the area of the haunch is equal to the areas of the individual ribs ΣA_{rib} over the effective width w_{eff} of the section. Therefore, the mean width of the haunch b_{haunch} can be calculated as

$$\left(b_{haunch}\right)_{\theta=0} = \left(\left(b_{eff}\right)_l + \left(b_{eff}\right)_r\right)\frac{A_{rib}}{A_{rib} + A_{void}} \qquad (2.10)$$

　　Unless the haunch is very deep, it can be assumed to have vertical sides instead of the sloping sides shown in Figure 2.4(b).

　　When $\theta = 90°$ in Figure 2.5, the ribs are transverse to the direction of the span of the composite beam. If we use the weakest cross-section in the analysis, this occurs at a section through a void between the ribs as shown in Figure 2.6. Hence in this case

$$\left(b_{haunch}\right)_{\theta=90°} = 0 \qquad (2.11)$$

　　The profile ribs may be oblique to the span of the beam, as shown in Figure 2.5 where $0 \le \theta \le 90°$. Hence, the effective cross-sectional shape lies between that shown in Figure 2.4(b) and Figure 2.6, and the effective width of the haunch lies between zero and $(b_{haunch})_{\theta=0}$. In general, the area of the haunch has only a minor effect on the

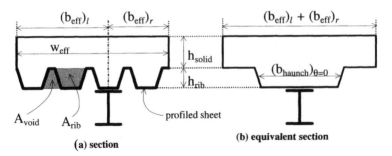

Figure 2.4 Longitudinal ribs

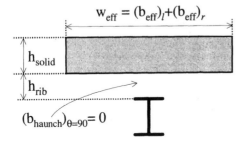

Figure 2.5 Oblique ribs

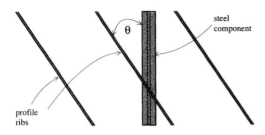

Figure 2.6 Transverse ribs

strength of the member. If, for example, the neutral axis under positive bending lies in the solid portion of the slab h_{solid} in both Figures 2.4(b) and 2.6, then the haunch is in tension and does not contribute to the strength of the beam as the tensile strength of the concrete is generally ignored. In addition, even if the neutral axis lies within the haunch, the difference in the flexural strength using different values of the haunch will only be small, since the haunch lies very close to the neutral axis of the section. Hence any reasonable variation between the extremes of Eqs. 2.10 and 2.11 may be used. For simplicity, we will assume here that the variation is linear, that is

$$\left(b_{haunch}\right)_\theta = \left(1 - \frac{\theta^o}{90}\right)\left(b_{haunch}\right)_{\theta=0} \tag{2.12}$$

2.3 Local buckling

2.3.1 *General*

The second effect that determines the size of a composite member is the *local buckling* response of the steel component. Local buckling occurs in thin-walled steel sections[4], and takes place when the steel section forms a more favourable or stable equilibrium position by buckling or distorting. A typical local buckle in the negative region of a composite beam is shown in Figure 1.20. The distortions shown in this figure vary along the length of the member, but local buckling is characterized by the flange-web junction remaining straight throughout the member's length. This is in direct contrast to lateral-distortional buckling that is treated in Chapter 13 as shown in Figure 1.19(a). Local buckling is precipitated by compressive stresses that arise from bending and axial compressive actions, and by shear. It is not caused by tensile stresses.

2.3.2 *Initial local buckling*

2.3.2.1 General

In a steel plate or a plate assembly such as an I-section component of a composite beam, the maximum stress to cause elastic buckling σ_{cr} is[4]

$$\sigma_{cr} = k\,\frac{\pi^2 E_s}{12\left(1 - v_s^2\right)}\,\frac{1}{\left(b/t\right)^2} \tag{2.13}$$

where E_s is the Young's modulus of the steel and v_s is its Poisson's ratio. In Eq. 2.13, b and t are the breadth and thickness of the plate element respectively, and k is the so-called local buckling coefficient. The local buckling coefficient depends on a number of factors, including the arrangement of loading and the restraint of the plate in a steel section. Values of k have been tabulated, and are given in standard texts[5].

2.3.2.2 **Example 2.3** *Plate slenderness limits in bending*

In the web element of an I-section member that is subjected to pure bending and which is simply supported at its connections to the flanges, the local buckling coefficient[5] k = 23.9 in Eq. 2.13 and where b = depth of the web d_w and t is the plate thickness of the web t_w. We will determine the plate slenderness b/t if the plate is to yield before buckling locally, that is its yield stress f_y is less than the local buckling stress σ_{cr} in Eq. 2.13. If E_s = 200 kN/mm² and v_s = 0.3 (see Section 1.3.2) are substituted into Eq. 2.13, then

$$\sigma_{cr} = \frac{23.9 \times \pi^2 \times 200 \times 10^3}{12 \times \left(1 - 0.3^2\right) \times \left(b/t\right)^2} \geq f_y$$

which may be rearranged to produce $(b/t)\sqrt{f_y/250} \leq 131$ where the units of f_y are in N/mm², and the normalizing of f_y by dividing by 250 N/mm² is a convenient manipulation to give the (b/t) limit some transparency, since yield stresses are often of Grade 250 (250 N/mm²). Hence for Grade 250 steel, provided the web is proportioned such that $(b/t) \leq 131$, we can develop the full yield stress before buckling occurs. It is worth noting that for higher strength steels ($f_y > 250$ N/mm²), the limit on (b/t) must actually drop below 131, since the elastic range before buckling is higher as yielding is delayed. It is worth noting that the term $(b/t)\sqrt{f_y/250}$ is often referred to as the modified slenderness as it is the slenderness b/t modified to take into account the yield strength f_y.

2.3.2.3 **Example 2.4** *Plate slenderness limits in shear*

Consider now the web of an I-section member that is subjected to pure shear and which is simply supported along its edges. The local buckling coefficient[5] for a long web without stiffeners in shear is k = 5.35. Hence if the plate yields in shear at $f_y/\sqrt{3}$ (Section 1.3.2) before buckling elastically at σ_{cr}, then

$$\sigma_{cr} = \frac{5.35 \times \pi^2 \times 200 \times 10^3}{12 \times \left(1 - 0.3^2\right) \times \left(b/t\right)^2} \geq \frac{f_y}{\sqrt{3}}$$

which may be rearranged to produce $(b/t)\sqrt{f_y/250} \leq 82$, where again the units of f_y are in N/mm².

2.3.3 *Section classifications*

2.3.3.1 General

When a composite cross-section is analysed, it is important to ensure that the steel component does not buckle locally, and this forms the basis of proportioning the cross-section in such a way that the desired limit state such as full plastification (Chapters 4 and 12) or first yielding (Chapter 3) occurs before the onset of local buckling. The types of cross-sections which correspond to the various limit states are called *plastic*, *compact*, *semi-compact* and *slender*. The rationale behind classifying a cross-section was illustrated in Examples 2.3 and 2.4, and this will be considered in the following in more detail.

Local buckling is more likely to take place in the negative or hogging moment regions than in the positive or sagging moment regions. When the neutral axis lies in the concrete component of a composite T-beam in positive bending, the steel is subjected only to tension and will not in general buckle, although local buckling is theoretically possible in regions of high shear. Buckling in shear is considered in Section 2.3.5.

2.3.3.2 *Plastic sections*

Plastic cross-sections are defined as those which allow a plastic mechanism to develop. Plastic mechanisms are considered in detail in Chapter 12, but we will note here that such cross-sections must be able to reach their strain hardening ranges (Section 1.3.2) before local buckling occurs. In addition, such sections must allow enough rotation in the strain hardening region for moment redistribution. This means that the depth to thickness ratios (b/t) for the elements of the steel component (the web and flanges) are restricted to quite low values, far more so for a web than Example 2.3 would suggest.

Figure 2.7 shows a composite T-beam in negative bending. The beam may be welded or hot-rolled, and there are different (b/t) ratios for both forms of fabrication owing to imperfections that are induced during their manufacture. For example, for a plastic section built-up by *welding*, the width to thickness ratio of the flange outstand (b_f/t_f) must satisfy

$$\frac{b_f}{t_f}\sqrt{\frac{f_y}{250}} \leq 8 \qquad\qquad (2.14)$$

while for a *hot-rolled* section, this ratio (b_f/t_f) must satisfy

$$\frac{b_f}{t_f}\sqrt{\frac{f_y}{250}} \leq 9 \qquad\qquad (2.15)$$

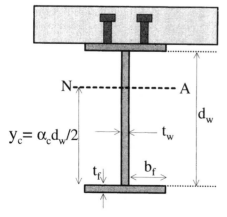

Figure 2.7 T-beam in negative bending

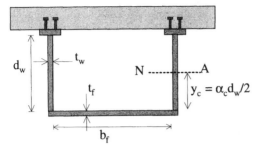

Figure 2.8 Box beam in negative bending

If the flange is not an outstand, but is simply supported as in the bottom flange of a box section as illustrated in Figure 2.8, the local buckling coefficient is much higher, so that the corresponding limits for a *welded* member are

$$\frac{b_f}{t_f}\sqrt{\frac{f_y}{250}} \leq 24 \qquad (2.16)$$

and for a *hot-rolled* member that would occur in a rolled box profile are

$$\frac{b_f}{t_f}\sqrt{\frac{f_y}{250}} \leq 27 \qquad (2.17)$$

The limiting depth to thickness ratio of the web (d_w/t_w) in Figure 2.8 depends on the amount of the web element subjected to compressive stresses. This can be found as a function of the plastic neutral axis parameter α_c shown in Figures 2.7 and 2.8, and defined as

$$\alpha_c = \frac{y_c}{d_w/2} \qquad (2.18)$$

where y_c is the depth of the plastic neutral axis from the inside of the compression flange of the steel component as shown in Figures 2.7 and 2.8. When the composite beam is subjected to negative bending, then for a plastic section classification it is suggested that the web depth to thickness ratio satisfies the inequality

$$\frac{d_w}{t_w}\sqrt{\frac{f_y}{250}} \leq \frac{82}{0.4 + 0.6\alpha_c} \qquad (2.19)$$

which is applicable to both welded and hot-rolled sections.

It is worth reiterating that, to achieve a plastic section as the yield stress f_y increases, a narrower or shallower plate must be chosen for a given thickness owing to the reciprocal relationship between the slenderness and $\sqrt{f_y}$. This is not paradoxical as the strength of the cross-section is related to f_y linearly and thus the section strength increases as the yield stress increases.

2.3.3.3 **Example 2.5** *Checking a plastic T-beam*

The composite T-beam with a welded steel component of yield strength $f_y = 250$ N/mm^2 shown in Figure 2.7 has the dimensions $d_w = 344$ mm, $b_f = 75$ mm, $t_w = 10$ mm and $t_f = 18$ mm. Suppose a rigid plastic analysis in negative bending (Chapters 4 and 12) indicates that the plastic neutral axis lies 204 mm above the inside of the bottom flange. Hence $(b_f/t_f)\sqrt{(f_y/250)} = (75/18)\sqrt{(250/250)} = 4.17 < 8$ in Eq. 2.14 and so the bottom flange is plastic. Also, $\alpha_c = 204/(344/2) = 1.19$, so that the limit in Eq. 2.19 is $82/(0.4 + 0.6 \times 1.19) = 74$ and $(d_w/t_w)\sqrt{(f_y/250)} = (344/10)\sqrt{(250/250)} = 34.4 < 74$ so that the web is also plastic. This section is thus suitable for plastic design.

2.3.3.4 Compact sections

A section is classified as compact if it buckles into the strain hardening region with sufficient rotation capacity to sustain the plastic moment, but may buckle locally before a full plastic mechanism (Chapter 12) may develop. Because of this, the limiting width to thickness ratios or depth to thickness ratios are relaxed slightly, and of course a plastic section will also be compact.

For a flange outstand built-up by *welding*, the limit is

$$\frac{b_f}{t_f}\sqrt{\frac{f_y}{250}} \leq 8.5 \tag{2.20}$$

and if the flange is *hot-rolled*, the limit is

$$\frac{b_f}{t_f}\sqrt{\frac{f_y}{250}} \leq 10 \tag{2.21}$$

In addition, for a simply supported *welded* plate (as in Figure 2.8) the limit is

$$\frac{b_f}{t_f}\sqrt{\frac{f_y}{250}} \leq 26 \tag{2.22}$$

and if the simply supported flange plate is produced by *hot-rolling*, then

$$\frac{b_f}{t_f}\sqrt{\frac{f_y}{250}} \leq 34 \tag{2.23}$$

A suggested limit on the web slenderness for the compact section classification is

$$\frac{d_w}{t_w}\sqrt{\frac{f_y}{250}} \le \frac{103}{\alpha_c} \qquad (2.24)$$

where α_c is defined in Figures 2.7 and 2.8, and in which Eq. 2.24 is assumed to be applicable to both welded and hot-rolled sections.

2.3.3.5 Semi-compact sections

Semi-compact sections allow the first yield moment $(M_s)_y$ of the steel component of a composite beam to develop, but the steel component may buckle locally before the full plastic moment $(M_s)_p$ is developed. The limits on the plate element slendernesses for the semi-compact classification are relaxed above those for compact sections, and of course both plastic and compact sections will satisfy the semi-compact classification.

For a *welded* flange outstand, the limit is

$$\frac{b_f}{t_f}\sqrt{\frac{f_y}{250}} \le 14 \qquad (2.25)$$

while for a *hot-rolled* flange outstand, the limit is

$$\frac{b_f}{t_f}\sqrt{\frac{f_y}{250}} \le 16 \qquad (2.26)$$

For a *welded* supported flange plate as in Figure 2.8, the limit is

$$\frac{b_f}{t_f}\sqrt{\frac{f_y}{250}} \le 30 \qquad (2.27)$$

while for a *hot-rolled* supported flange plate as in Figure 2.8, the limit is

$$\frac{b_f}{t_f}\sqrt{\frac{f_y}{250}} \le 41 \qquad (2.28)$$

As for a compact section, the limiting web depth to thickness ratio is not dependent on whether it is hot-rolled or welded, and a suggested limit is

$$\frac{d_w}{t_w}\sqrt{\frac{f_y}{250}} \le \frac{115}{\alpha_c} \qquad (2.29)$$

where α_c in Figures 2.7 and 2.8 is obtained from an *elastic* analysis based on transformed areas, as described in Chapter 3. If there was no reinforcement in the slab and the steel component was a doubly symmetric I-section, the neutral axis would be at the mid-height of the steel component and α_c would be 1.0. The limit in Eq. 2.29 would then be 115, which is lower than the limit of 131 calculated in Example 2.3 that ignored residual stresses.

2.3.3.6 **Example 2.6** *Section strength of a semi-compact section*

If we define M_s to be the cross-sectional strength of the steel component, then it was shown that for plastic and compact sections, $M_s = (M_s)_p$ which is the full plastic moment of the steel component. Note that in negative bending, the section capacity of the composite section will be greater than this, owing to the presence of tensile forces in the reinforcement in the cracked concrete component which also contributes to the bending capacity.

Consider now a welded box section composite beam, as in Figure 2.8, whose webs can be considered as being compact, but with a flange modified slenderness $(b_f/t_f)\sqrt{(f_y/250)} = 27.5$. Suppose the moment to cause first yield of the steel component is $(M_s)_y = 140$ kNm and that to cause full plasticity of the steel component $(M_s)_p = 155$ kNm. Clearly because the modified flange slenderness of 27.5 is less than the limit of 30 in Eq. 2.27, the section can be considered as semi-compact, but because it is greater than the limit of 26 in Eq. 2.22 it is not compact. This means that a moment in the steel component of 140 kNm is attainable, but that the plastic moment of 155 kNm is not. We can, however, interpolate linearly between these two moments based on the value of the section slenderness. In this case, the capacity of the steel section $M_s = 140 + (155 - 140) \times (30 - 27.5)/(30 - 26) = 149.4$ kNm. This increase of 7% above the first yield moment of the steel component should not be ignored in the analysis of a semi-compact section, and the increase for T-section beams may be much higher than this.

It can thus be seen that for plastic and compact sections, the section strength of the steel component in negative bending $M_s = (M_s)_p$, while for semi-compact sections $(M_s)_p \geq M_s \geq (M_s)_y$. Of course, semi-compact sections are unsuitable for the rigid plastic analysis techniques of Chapter 4, which are restricted to plastic and compact sections. Semi-compact sections must be analysed by the linear elastic techniques of Chapter 3.

It is worth reiterating that the moments $(M_s)_y$ and $(M_s)_p$ are the moments in the steel component when a moment M_a is applied to the composite section. For sections that remain elastic, the moment in the steel component when the composite member just starts to yield $(M_s)_y$ can be determined from an elastic analysis of the section as described in Chapter 3. For example, if a moment $(M_a)_y$ is applied to a composite section of flexural rigidity $(EI)_{cmp}$ to cause yield, then the curvature in the composite section is $\kappa_y = (M_a)_y/(EI)_{cmp}$ which is also the curvature in the steel component. Hence $(M_s)_y = \kappa_y (EI)_s$ where $(EI)_s$ is the flexural rigidity of the steel component.

The moment in the steel component when the composite section is fully plastic $(M_s)_p$ can be determined from the distribution of stresses in the steel component as shown in Figures 1.16(b), (e) and (h) (the method for determining these stress distributions is described in detail in Chapter 4). For example let us consider the stress distribution in Case 2 at (e) in Figure 1.16 which is shown again in Figure 2.9(b). It is only necessary to consider the stress distribution in the steel component

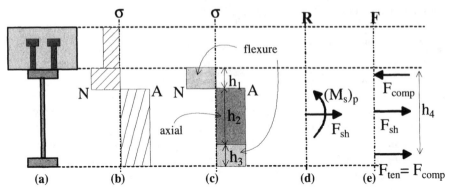

Figure 2.9 Moment in steel component

which is shown by itself in (c). This stress distribution resists the axial load F_{sh} and moment $(M_s)_p$ shown in (d). Hence the stress distribution in (c) can now be visualized as that required to resist flexure over regions h_1 and h_3 and that required to resist the axial load over region h_2 as shown. As the neutral axis position N-A is already fixed by a rigid plastic analysis from Chapter 4, h_1 is fixed so that the magnitude and position of the axial force above the neutral axis, F_{comp} in (e), can be determined directly. The compressive force F_{comp} is in equilibrium with a tensile force at the bottom of the steel component such that $F_{ten} = F_{comp}$, hence the position of F_{ten} can be determined from which $(M_s)_p = F_{comp}h_4$. The remaining stress distribution in (c) over region h_2 simply resists the longitudinal shear in the composite beam F_{sh}.

2.3.3.7 Slender sections

Slender sections possess plate element width to thickness ratios that exceed those for the semi-compact classification, and their steel components buckle at moments below that to cause first yield $(M_s)_y$. Slender composite cross-sections occur in deep T-beam bridge section girders, or in slender box sections.

There are two ways of determining the strength of slender sections. The first is to use an effective width approach, similar to that in Section 2.2 for shear lag, in which regions beyond the effective area are ignored. However, this generally renders the section monosymmetric, and the calculation of the section properties is quite involved. The basis for the effective width approach is the post-local buckling response of thin steel plates[4,5].

The second way is to simply factor the moment at first yield by the ratio of the limit for a semi-compact section to the actual modified slenderness ratio (b_f/t_f) $\sqrt{(f_y/250)}$ or $(d_w/t_w)\sqrt{(f_y/250)}$ for the section.

2.3.3.8 **Example 2.7** *Section strength of a slender section*

Consider the steel component of a welded composite T-beam that has a moment at first yield, calculated on the full steel cross-section of $(M_s)_y = 120$ kNm, and which has a value of α_c based on an elastic analysis of 1.25. Suppose the web has a modified slenderness ratio of $(d_w/t_w)\sqrt{(f_y/250)} = 105$ and the modified slenderness ratio of the flange is 19. The limit in Eq. 2.29 is $115/1.25 = 92 < 105$ and so the web is slender. Moreover, the limit in Eq. 2.25 is $14 < 19$ and so the flange is slender also. It is worth noting that only one of the steel elements (flange outstands or web) being slender renders the section classification as being slender. The section strength of the steel component may then be calculated as $M_s = (92/105) \times (14/19) \times 120 = 80.0$ kNm. A section strength for the steel element of 80.0 kNm would then be used in a linear elastic analysis. Note that the method of this example, although very simple, is conservative.

2.3.4 *Beams in positive bending*

Beams in positive bending have the neutral axis in the web above the centroid of the steel component. These web elements are not generally prone to local buckling, since at their connection to the top flange they are restrained by the attachment to a rigid concrete component, while at the neutral axis level they are restrained by the tensile portion of the steel component. We may therefore consider the webs as being plastic, although in deep beams this classification may be unconservative, and should be viewed with care.

When the neutral axis lies in the steel web, the top steel flange will be subjected to compression. If the section is to considered plastic or compact, as described in the previous paragraph, the top steel flange must not buckle away from the concrete slab, as shown in Figure 2.10. It is suggested that to prevent this buckling, the spacing s of the stud shear connectors should not exceed

$$\frac{s}{t_f}\sqrt{\frac{f_y}{250}} \leq 30 \tag{2.30}$$

2.3.5 *Local buckling in shear*
2.3.5.1 Slenderness limit

It will be shown in Chapters 4 and 9 that it is usual to assume that all the shear carried by the cross-section of a composite beam is resisted by the web element

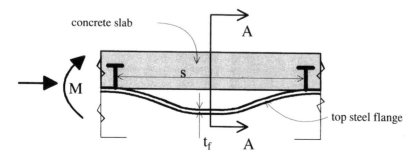

Figure 2.10 Buckling of top flange in positive bending

of the steel component. If the web element is proportioned so that it may develop its full yield strength, then it is said to be *stocky*, and the depth to thickness ratio must satisfy

$$\frac{d_w}{t_w}\sqrt{\frac{f_y}{250}} \leq 82 \tag{2.31}$$

Normally if this limit is exceeded and the web is *slender*, it is usual to stiffen the web with vertical web stiffeners. However, vertical web stiffeners in the steel component are beyond the scope of this book, and we can calculate the strength of the unstiffened web on the basis of initial local buckling as described in the following example.

2.3.5.2 **Example 2.8** *Section strength of a slender web in shear*

Suppose the web element of a steel component has a yield strength $f_y = 300$ N/mm² and has dimensions $d_w = 1000$ mm and $t_w = 8$ mm. The modified slenderness $(d_w/t_w)\sqrt{(f_y/250)} = (1000/8)\sqrt{(300/250)} = 136.9 > 82$ from Eq. 2.31 and so the web is slender in shear. If we note that the yield strength of the web in shear is $(d_w t_w)\tau_y/\sqrt{3}$ $= 1000 \times 8 \times 300/\sqrt{3} = 1386$ kN (Section 1.3.2), then its section strength may simply be determined from $(82/136.9)^2 \times 1386 = 497$ kN as the buckling strength from Eq. 2.13 is inversely proportional to $(b/t)^2$. The strength of the slender web is only 36% of its yield strength, and a more economic design would involve the use of a thicker web or the use of vertical web stiffeners.

2.3.6 *Concrete-filled steel tubes*

Concrete-filled steel tubes find widespread use as compression members, and their flexural buckling is considered in Chapter 8. The section classifications for a compression member are either *slender* or *semi-compact*, since the compact and plastic moments are irrelevant in the absence of bending because the member is subjected to compression throughout. A semi-compact section allows yielding and hence the squash load of the steel component of the tube to be attained before local buckling, while slender sections buckle locally before the yield stress is reached, and so the steel component of slender compression members buckles locally before it reaches its squash load.

Unlike hollow steel rectangular sections, local buckling of concrete-filled steel tubes is resisted by the restraint provided by the concrete core, as shown in Figure 2.11. It is suggested that for a concrete-filled tube fabricated by *welding*, the b/t limit shown in Figure 2.11 should satisfy

$$\frac{b}{t}\sqrt{\frac{f_y}{250}} \leq 47 \tag{2.32}$$

while if it is fabricated by *hot-rolling*, the b/t limit should satisfy

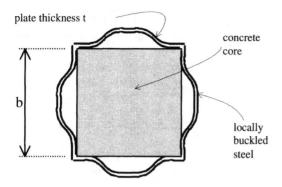

Figure 2.11 Local buckling of a concrete-filled tube

$$\frac{b}{t}\sqrt{\frac{f_y}{250}} \le 66 \qquad (2.33)$$

It can be seen that the above two limits are much greater than those for a supported flange (30 and 41 for a welded and hot-rolled flange plate respectively, as in Eqs. 2.27 and 2.28) owing to the restraint provided by the concrete core.

If the composite column is a concrete-filled circular steel tube of outside diameter d_o and thickness t, then the slenderness d_o/t should satisfy

$$\frac{d_o}{t}\left(\frac{f_y}{250}\right) \le 82 \qquad (2.34)$$

The provision of Eq. 2.34 is based on that generally used for a hollow circular tube, and in the absence of reliable data for concrete-filled circular tubes this limit is conservative.

Of course, an encased steel section as shown in Figure 1.4(a) is unlikely to buckle locally, unless the concrete cover is very thin. Occasionally, however, only the void between the steel flanges is filled with concrete, and these flanges form partial and permanent formwork for the concrete as in Figure 2.12. In this case, the web is

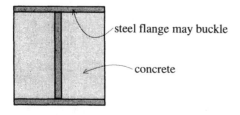

Figure 2.12 Partially encased I-section

prevented from buckling locally, but the flanges must satisfy the limits of Eqs. 2.25 and 2.26 if they are to remain fully effective and not buckle.

2.4 References

1. Timoshenko, S.P. and Goodier, J.N. (1970). Theory of Elasticity. 3rd edn, McGraw Hill, New York.
2. Eurocode 4 (1994). Design of Composite Steel and Concrete Structures. DDENV 1994-1-1, Draft for development.
3. Ansourian, P. (1975). 'An application of the method of finite elements to the analysis of composite floor systems'. Proceedings of the Institution of Civil Engineers, London, Part 2, Vol. 59, 699-726.
4. Trahair, N.S. and Bradford, M.A. (1998). The Behaviour and Design of Steel Structures to AS4100. 3rd edn, E&FN Spon, London.
5. Bulson, P.S. (1970). The Stability of Flat Plates. Chatto and Windus, London.

3 Elastic analysis of composite beams

3.1 Introduction

In ultimate strength analyses, such as the rigid plastic analysis of Chapter 4, the behaviour of the composite member is governed essentially by inelasticity and the nonlinear behaviour of the steel and concrete components. However, in their day to day life, composite structures are usually loaded well below levels that would cause failure, and the behaviour of the steel, concrete and shear connection can be considered as *linear*. In limit states terminology, we refer to behaviour at these lower load levels as service load behaviour, or to the *serviceability limit state*.

Service loads are the loads usually experienced by a member over a relatively long period of time, including self weight and sustained loads, and by short-term lower level live loads. Satisfaction of the serviceability limit state is important, as it must be ensured that the composite structure does not deflect excessively, that is does not vibrate greatly and that crack widths in the concrete component remain sufficiently small. Analyses to guard against the attaining of these serviceability limit states are based on *linear elastic assumptions*, rather than the plastic assumptions of Chapter 4. It is worth noting that fatigue design, which is treated in Chapters 14 and 15, is carried out using linear elastic analysis, even though fatigue is a failure criterion. This is also true for lateral-distortional buckling treated in Chapter 13, where linear elastic analysis again is used in the prediction of a strength failure mode.

3.2 Linear material properties

In order to undertake an elastic analysis, we must assume that the relationship between stress and strain, or load and deformation, is linear for the steel and concrete components, as well as for the reinforcement and the shear connectors. The material properties for the steel, concrete and reinforcement were described fully in Chapter 1. Stud shear connectors are treated in detail in Chapter 5, but we only need to note here that their response is linear elastic for a substantial range of loads, with the ratio of the shear force to the corresponding shear deformation being expressed by the stiffness or modulus K. In terms of the degree of interaction introduced in Section 1.5.2, a full interaction analysis (as considered in Section 3.3) is characterized by $K \to \infty$ in which there is no slip and hence no slip strain at the steel/concrete interface as in Figure 1.18(b), and by $K = 0$ when there is no interaction between the steel and concrete components, so that the interface may be considered as greased and hence the slip strain is at its maximum as shown in Figure 1.18(a). The condition of partial interaction is therefore dependent on a finite value of K, but because of the complexities that arise in such a condition, we shall concentrate here on the condition of full interaction, that is when $K \to \infty$.

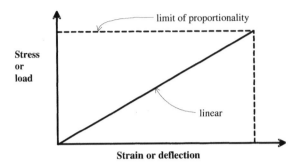

Stress or load / Strain or deflection / limit of proportionality / linear

Figure 3.1 Linear elastic material properties

The schematic representation of the linear elastic range of the component materials is shown in Figure 3.1. The ratio of stress to strain in the steel, concrete and reinforcement is constant. For the steel component, it is assumed that this ratio is equal to $E_s = 200$ kN/mm² in tension and compression until the yield stress f_y is reached, and for the reinforcement that, again, $E_s = 200$ kN/mm² up to the yield stress f_{yr}. For the concrete in the short-term, the stress to strain ratio is assumed to be governed by E_c in Eq. 1.4 and in the long-term by E_e in Eq. 1.7. Because the stress-strain response of the concrete becomes nonlinear well before its compressive strength f_c is reached, it is usual practice to assume linear elastic behaviour is governed by the moduli E_c or E_e up to about 40% or 50% of f_c.

The mechanical response of shear connectors is usually stated in terms of the load-slip behaviour, as noted earlier. The modulus K (whose units are force/length) may usually be considered as constant for stud shear connectors loaded statically up to about 70% of their dowel strength D_{max}. This is again depicted in Figure 3.1.

3.3 Full interaction analysis

3.3.1 *Elastic transformed cross-sections*

3.3.1.1 Assumptions

For full interaction, it is assumed that the slip and hence slip strain at the steel/concrete interface are negligible, that is $K = \infty$. Furthermore, under service loading the short-term and effective concrete moduli E_c and E_e respectively will be considered constant as noted in the previous section, and this forms the basis for transformed area analysis. As is usual in elastic structural mechanics, the cross-section is transformed into an equivalent concrete section according to the modular ratios $n = E_s/E_c$ for short-term loading, and $n_e = E_s/E_e$ for long-term loading.

3.3.1.2 **Example 3.1** *Transformed cross-sections in positive bending*

Consider the composite beam shown in Figure 3.2 that is subjected to positive bending. For short-term behaviour, the modular ratio is $n = 200/26.8 = 7.0$. The composite section can be transformed into an equivalent concrete section by

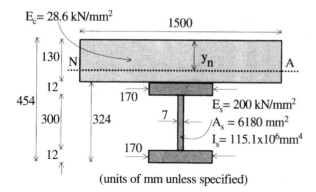

(units of mm unless specified)

Figure 3.2 Composite steel and concrete beam

increasing the area of the reinforcement to $(n - 1)A_r$ (and using the gross concrete area) and the area of the steel component to nA_s as shown in Figure 3.3. In this example, there is no reinforcement and the transformed area of steel is $7.0 \times 6180 = 43,260$ mm². The transformation process is shown in Figure 3.3. Note also that the transformed second moment of area of the steel is $nI_s = 7.0 \times 115.1 \times 10^6 = 806 \times 10^6$ mm⁴ with the same steel depth of 324 mm.

If the concrete is assumed to be uncracked, then the neutral axis will lie at the centroid of the transformed section. By taking first moments of area about the top fibre, $(130 \times 1500 + 43,260) y_n = 1500 \times 130 \times (130/2) + 43,260 \times (130 + 324/2)$, and so the depth of the neutral axis below the top fibre is $y_n = 106.2$ mm. The transformed second moment of area about the neutral axis is $I_{nc} = 130^3 \times 1500/12 + 130 \times 1500 \times (130/2 - 106.2)^2 + 806 \times 10^6 + 43,260 \times (130 + 324/2 - 106.2)^2 = 2905 \times 10^6$ mm⁴.

Since the depth to the neutral axis is 106.2 mm < 130 mm, the neutral axis lies in the concrete component as in Figure 3.3(b) and the concrete below this axis is

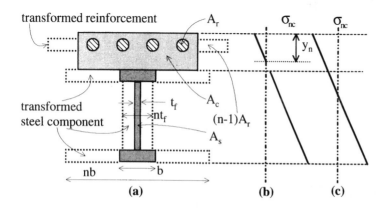

Figure 3.3 Composite beam transformed to a concrete beam

subjected to tension. Although the concrete does have a small tensile strength (Section 1.3.5.1), it is usual to ignore this and assume that its tensile strength is zero. The position of the neutral axis in the cracked section can then be determined by equating the first moment of area about the neutral axis of the transformed areas above and below it. Hence $1500y_n^2/2 = 43,260 \times (130 + 324/2 - y_n)$ from which $y_n = 104.1$ mm, and the second moment of area of the uncracked transformed section about this axis is $I'_{nc} = 2897 \times 10^6$ mm^4. The flexural rigidity of the cracked section is thus only 0.3% less than that of the uncracked section as the cracked region is adjacent to the neutral axis, and hence cracking can be ignored in standard composite T-sections in positive bending. It is also worth noting that making the steel beam composite with the concrete slab has increased the flexural rigidity of the member by E_cI_{nc}/E_sI_s or 360%, so that the deflection of the composite beam will be only 28% of the steel beam of the same length when acting by itself.

3.3.1.3 **Example 3.2** *First yield of a composite section in positive bending*

The analysis in Example 3.1 is valid while the steel and concrete remain elastic. Let us suppose that the section remains linearly elastic until the bottom fibre of the steel yields at $f_y = 300$ N/mm^2. The strain at this level is $\varepsilon_y = 300/200 \times 10^3 = 1.5 \times 10^{-3}$. Hence assuming a full interaction analysis, that is that there is no slip strain at the steel/concrete interface, the strain at the top fibre of the concrete is $(106.2/(454 - 106.2)) \times 1.5 \times 10^{-3} = 0.458 \times 10^{-3}$. This top fibre concrete strain produces a concrete stress of $28.6 \times 10^3 \times 0.458 \times 10^{-3} = 13.1$ N/mm^2. This concrete stress is in the elastic range, as the concrete component would generally have a strength $f_c \geq 25$ N/mm^2, and the maximum concrete stress at first yield of the steel is only about half of this concrete strength. It is usually the case that elastic analysis of a T-section remains valid up to first yield of the steel.

The moment of resistance of the composite section at first yield may also be calculated easily. The curvature at first yield is $\kappa_y = 1.5 \times 10^{-3}/(454 - 106.2) = 4.31 \times 10^{-6}$ mm^{-1}, and so the first yield moment $M_y = (E_cI_{nc})\kappa_y = 28.6 \times 10^3 \times 2905 \times 10^6 \times 4.31 \times 10^{-6}$ Nmm $= 358.1$ kNm. If the steel component acted alone, the moment in the steel beam to cause first yield is $M_y = f_yI_s/(h_s/2) = 300 \times 115.1 \times 10^6/(324/2)$ Nmm $= 213.1$ kNm. The first yield moment of the composite beam is thus 168% of that of the steel component alone.

The steel component for this beam subjected to positive bending is in tension throughout and therefore there is no need to check the section classifications of Chapter 2 (which require at least a semi-compact steel section). For the situation when the top region of the web is subjected to compression as in Figure 3.3(c), the web is restrained rigidly against buckling by its attachment to the top flange, which is connected by the shear connectors to the concrete, and the lower part of the compression region of the web is restrained by its tensile portion. The web is therefore unlikely to buckle, and a check of its section classification is not generally needed.

3.3.1.4   **Example 3.3** *Transformed cross-sections in negative bending*

Let us assume that the cross-section in Figure 3.2 is now subjected to a negative moment. Unless the neutral axis lies in the concrete component, the concrete will have little or no effect as it is subjected to tension, and the strength of the bare steel beam alone must be enhanced by providing reinforcement in the slab. We will assume the concrete component has 0.6% reinforcement ($A_r = 1170$ mm^2) that has the same elastic modulus as the steel component, and that this reinforcement is positioned 35 mm below the top surface of the slab.

If it is first assumed that the neutral axis lies within the steel component, then the composite beam will consist of the steel component in compression and tension and the reinforcement component in tension. As both of these components have the same elastic modulus, E_s, there is no need to transform the beam and it can be analysed as a steel beam. Performing this analysis, by taking first moments of areas about the lower fibre of the bottom flange leads to the neutral axis being $y_n = 203$ mm from the bottom fibre of the steel component, which is well within the steel. The second moment of area of this 'steel' section is therefore $I_{ns} = 115.1 \times 10^6 + 6180 \times (324/2 - 203)^2 + 1170 \times (454 - 35 - 203)^2 = 180.1 \times 10^6$ mm^4. Even though the concrete is ineffective in flexure, the flexural rigidity of the steel beam-reinforcement component is 56% greater than that of the steel beam by itself, and $180.1 \times 7/2905$ or 43% of the flexural rigidity of the composite section in positive bending that was determined in Example 3.1.

If the composite beam has a very large area of reinforcement or a very deep slab, then the neutral axis may lie in the concrete component. In this case the area of concrete below the neutral axis is uncracked, and the position of the neutral axis can be found by equating the first moment of area about the neutral axis of the transformed section above to that of the transformed section below, in a similar fashion to Example 3.1. It is worth reiterating that the likelihood of the neutral axis being in the concrete portion of a standard composite T-beam is remote.

3.3.1.5   **Example 3.4** *First yield of a composite section in negative bending*

An illustration similar to that of Example 3.2 will be used to study the behaviour of the beam in Figure 3.2 when subjected to negative bending. Again, let us suppose the behaviour is linearly elastic until first yield of the bottom fibre of the steel component in compression at a stress of $f_y = 300$ N/mm^2. The curvature is then $\kappa_y = 300/(200 \times 10^3 \times 203) = 7.39 \times 10^{-6}$ mm^{-1}. The tensile strain in the reinforcement is thus $7.39 \times 10^{-6} \times (454 - 35 - 203) = 1.60 \times 10^{-3}$ and the stress is $200 \times 10^3 \times 1.60 \times 10^{-3} = 320$ N/mm^2. This stress is below the typical yield stress of $f_{yr} = 400$ N/mm^2 for the reinforcement, and indeed in the majority of cases the reinforcement remains elastic until first yield of the steel component is reached. If in fact the reinforcement has yielded, the bending capacity will be slightly overestimated if elastic procedures are used, but the error in linear elastic analysis is generally minuscule.

At first yield, the moment capacity $M_y = E_s I_{ns} \kappa_y = 200 \times 10^3 \times 180.1 \times 10^6 \times 7.39 \times 10^{-6}$ Nmm = 266.2 kNm. This is 266.2/358.1 or 74% of the first yield moment in positive bending, and 266.2/213.1 or 25% greater than the first yield moment of the steel section alone.

For first yield to be achievable in negative or hogging bending, the steel component must be semi-compact. Hence for the flanges, $(b_f/t_f)\sqrt{(f_y/250)} = ((170 - 7)/(2 \times 12) \times \sqrt{(300/250)} = 7.4 < 14$ (Eq. 2.25), and so the flanges are compact. For the web, $\alpha_c = (203 - 12)/(300/2) = 1.27$ (Eq. 2.18), and so $\alpha_c(d_w/t_w) \sqrt{(f_y/250)} = 1.27 \times (300/7) \times \sqrt{(300/250)} = 59.8 < 115$ (Eq. 2.29) and so the web is also semi-compact.

3.3.2 *Continuous composite beams*

A continuous composite beam is shown in Figure 3.4. Within the lengths $(L_c)_l$ and $(L_c)_r$, the beam is subjected to positive bending and has a transformed flexural rigidity $E_c I_{nc}$, while in the region $(L_c)_c$ the beam is in negative bending and has a transformed flexural rigidity $E_s I_{ns}$. Because the positive (or sagging) flexural rigidity is greater than the negative (or hogging) flexural rigidity as shown in Examples 3.1 and 3.3, the response of the beam is that of a nonuniform or 'stepped' member. The difficulty with analysing continuous composite beams, even in the linear elastic range of structural response, is that the internal points of inflection are not known at the outset, so that an iterative scheme must be followed to determine the extent of the positive and negative bending regions.

In fact, the force method of structural analysis[1] may be used conveniently to analyse the two-span composite beam shown in Figure 3.4. Firstly, the positive bending flexural rigidity $E_c I_{nc}$ as determined by the method of Example 3.1 can be assumed throughout, and the force method used to calculate the redundant vertical reaction at B in the figure. This then allows the bending moment distribution M to be determined, and the region $(L_c)_c$ to be identified over the internal support. In the positive moment regions $(L_c)_l$ and $(L_c)_r$, the curvature is $M/E_c I_{nc}$, while in the negative moment region $(L_c)_c$ the curvature is $M/E_s I_{ns}$, where I_{ns} is determined as in Example 3.3 and M is the moment at any point along the beam. Of course, the curvatures in the positive and negative moment regions of the beam are of different sign, as the moment is positive in the sagging moment region and negative in the hogging moment

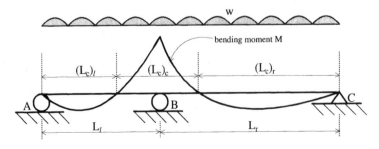

Figure 3.4 Two-span continuous beam

region. Using these curvatures, the force method may again be used to determine an updated estimate of the vertical reaction at B, so that an updated bending moment diagram may be drawn and updated estimates of the positive and negative bending regions established. The curvatures in these regions are again calculated for use in the force method, and a second estimate of the vertical reaction at the internal support B may be determined. In this hand method, which lends itself to computer programming, the vertical reaction at B is calculated iteratively until it converges to an acceptable tolerance. The final bending moment for the continuous beam may thus be established.

It is worth noting that such an analysis must be carried out to undertake a lateral-distortional buckling study of a continuous beam, as discussed in Chapter 13. This is because lateral-distortional buckling can cause failure of the steel component over the hogging region $(L_c)_c$. Note, too, that the analytical technique presented in the previous paragraph is based on linear elastic principles, although the solution strategy is iterative.

For a general composite beam continuous over a number of spans, a commercially available stiffness-based computer program may be used. The sagging and hogging flexural rigidities $E_c I_{nc}$ and $E_s I_{ns}$ respectively are determined as in Examples 3.1 and 3.3. For a given loading, the program is invoked using the positive or sagging rigidity $E_c I_{nc}$ throughout, and the bending moment diagram is drawn and the points of contraflexure identified. The program is again used using $E_c I_{nc}$ in the identified positive region and $E_s I_{ns}$ in the identified negative region, the bending moment diagram redrawn and the revised points of contraflexure identified. The analysis is undertaken iteratively by determining the positive and negative moment regions until the reactions or moments at the supports converge to a suitable accuracy. With graphics capabilities, this analysis may be undertaken with relative ease.

3.3.3   *Deflections due to creep*

3.3.3.1   General

By using simple modular ratio theory in a full interaction analysis, it is possible to determine relatively accurately the deflections of a composite beam caused by creep. This merely requires a transformation of the area according to the modular ratio $n_e = E_s/E_e$ instead of the modular ratio $n = E_s/E_c$.

3.3.3.2   **Example 3.5** *Deformations induced by creep*

The beam considered in Example 3.1 spans 6 m and is acted upon by a sustained uniformly distributed load of w = 45 kN/m. In Example 3.1, the transformed second moment of area I_{nc} was calculated to be 2897×10^6 mm^4. Under short-term or instantaneous loading, the deflection would then be $(5/384) \times (45 \times 6000^4)/(2897 \times 10^6 \times 28.6 \times 10^3) = 9.1$ mm.

Consider now long-term loading for which the creep coefficient $\phi = 3$. The effective modulus $E_e = 28.6/(1 + 3) = 7.15$ kN/mm^2 (Eq. 1.7) and the long-term modular ratio $n_e = 200/7.15 = 28$. This produces a transformed steel area of $n_e A_s = 173 \times 10^3$ mm^2 and a transformed steel second moment of area of $n_e I_s = 3233 \times 10^6$ mm^4. Because most of the transformed area now lies below the slab, we will assume that the depth to the neutral axis below the top fibre y_n lies in the steel. It can then be shown that y_n = 171.1 mm which indeed is 41.1 mm below the soffit of the slab. The transformed second moment of area I_{nc} can be calculated similarly to Example 3.1 to be 7964 × 10^6 mm^4. Using the value of E_e = 7.15 kN/mm^2 produces a long term deflection of 13.3 mm which is a 46% increase in deflection above the short-term value. It is worth noting that shrinkage can increase this time-deflection even more, and this effect is considered in the following sub-section.

3.3.4 *Deflections due to shrinkage*

3.3.4.1 Behaviour

Shrinkage is time-dependent, and therefore the forces that are induced will cause creep. Because of this, the effective modulus E_e introduced in Section 1.3.5.2 should be used in a shrinkage analysis.

The effect of shrinkage in the sagging region of a composite T-beam is shown in Figure 3.5. In the absence of shear connectors, the concrete will contract as shown in (a). The shear connectors resist this contraction as shown in (b), so that the shear forces on the connectors due to shrinkage oppose those due to gravity loads as indicated by the distorted shape of the connectors. However, contraction of the concrete through shrinkage will cause the beam deformation shown in Figure 3.5(b), and so induce deflections and flexural stresses that are in the same direction as those induced by gravity loads.

In order to quantify the forces and deformations induced by shrinkage, consider the right hand side of the beam shown in Figure 3.5(a) that does not have any shear connectors and which is also shown in Figure 3.6. The concrete component is allowed to contract due to shrinkage as shown, producing a lack of fit of $\varepsilon_{sh}L/2$, where ε_{sh} is the shrinkage strain, as shown in Figure 3.5(a). The shear connectors will resist this contraction as shown in (b), and therefore enforce compatibility. Again, following the linear elastic assumption of this chapter we will assume that there is zero slip, so that the analysis is a full interaction analysis based on transformed sections.

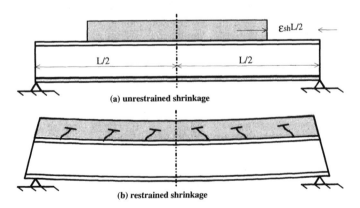

(a) unrestrained shrinkage

(b) restrained shrinkage

Figure 3.5 Shrinkage deformations

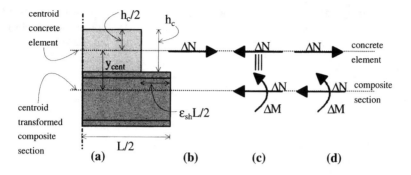

Figure 3.6 Equivalent force system for shrinkage

3.3.4.2   **Example 3.6** *Deformations caused by shrinkage*

Let us consider again the section in Example 3.1 that is 6 m long and has a long-term shrinkage strain of $\varepsilon_{sh} = 500 \times 10^{-6}$. In order to prevent slip and hence lack of fit, as shown in Figure 3.6(a), an axial force ΔN in (b) has to be applied to the concrete component. This axial force must lie at the centroid of the concrete component (ignoring any reinforcing as only the concrete is shrinking) and has a magnitude $\Delta N = E_e \varepsilon_{sh} A_c = 7.15 \times 10^3 \times 500 \times 10^{-6} \times 1500 \times 130$ N = 697.1 kN. When this value of ΔN is applied to the concrete component in (b) it ensures compatibility, but the system is now not in equilibrium. The net effect of maintaining equilibrium by applying an equal and opposite force ΔN in line with the concrete centroid is to apply an axial compression ΔN and a moment ΔM at the centroid of the composite section, as in (c). The system of forces shown in (d) is thus in equilibrium, and clearly $\Delta M = y_{cent} \Delta N = (171 - 130/2) \times 697.1$ kNmm = 73.96 kNm, which is the couple formed from the two forces ΔN.

It is thus clear that the effect of shrinkage is to produce a constant moment of 73.96 kNm in this case over the full length L of the beam. Such a constant moment will produce a deflection of $\Delta M L^2 / 8 E_e I_{nc} = 73.96 \times 10^6 \times 6000^2 / (8 \times 7.15 \times 10^3 \times 7964 \times 10^6) = 5.8$ mm. The total time-dependent deflection may then be approximated as $13.3 + 5.8 = 19.1$ mm.

It is worth noting that it has been shown[2] that relaxing the condition of full interaction does not influence greatly the time-dependent results, and that the above analyses techniques are satisfactory.

3.4   Partial shear connection

3.4.1   *Simplified model*

Linear elastic analyses utilizing partial interaction when $0 < K < \infty$ are complex. The concept of partial-shear-connection was introduced in Chapter 1, and obviously the degree of interaction, which influences the deflections, depends on the degree of shear connection η_{max}. Obviously when $\eta_{max} = 0$ there is no interaction as there are no shear connectors, and the deflection of the beam v_{no} depends only on the flexural rigidities $E_c I_c$ of the concrete component and $E_s I_s$ of the steel component.

It has been suggested[3] that the deflection under partial interaction v_{part} can be determined from that assuming full interaction v_{full} and no interaction v_{no} by the empirical equation

$$v_{part} = v_{full} + \alpha\left(v_{no} - v_{full}\right)\left(1 - \eta_{max}\right) \tag{3.1}$$

where the coefficient α is taken as 0.4, and where the flexural rigidity for no interaction is given by

$$\left(EI\right)_{no} = E_c I_c + E_s I_s \tag{3.2}$$

3.4.1.1 **Example 3.7** *Short-term deflections with partial shear connection*

We will again analyse the cross-section shown in Figure 3.2, and it will be assumed that this beam spans 6 m and is acted upon by a short-term uniformly distributed load of $w = 45$ kN/m. The flexural rigidity with no interaction $(EI)_{no} = 28.6 \times 10^3 \times 130^3 \times 1500/12 + 200 \times 10^3 \times 115.1 \times 10^6 = 3.09 \times 10^{13}$ Nmm^2. The deflection v_{no} is thus $(5/384) \times 45 \times 6000^4 / 3.09 \times 10^{13} = 24.6$ mm. From Example 3.5, $v_{full} = 9.1$ mm. Let us assume that the degree of shear connection is $\eta_{max} = 0.5$. Hence from Eq. 3.1, $v_{part} = 9.1 + 0.4 \times (24.6 - 9.1) \times (1 - 0.5) = 12.2$ mm. Decreasing the degree of shear connection from 100% used in full interaction analysis to 50% increases the deflection by 12.2/9.1 or 34%.

3.4.1.2   **Example 3.8** *Long-term deflection with partial shear connection*

Example 3.7 will now be reworked assuming partial interaction, except that long-term properties associated with $\phi = 3$ and $\varepsilon_{sh} = 500 \times 10^{-6}$ will be used as in Examples 3.5 and 3.6. Under a condition of no interaction, the value of E_e must be used, but the effects of shrinkage are of course irrelevant. Hence $(EI)_{no} = 7.15 \times 10^3 \times 130^3 \times 1500/ 12 + 200 \times 10^3 \times 115.1 \times 10^6 = 2.50 \times 10^{13}$ Nmm^2 and so the deflection $v_{no} = 30.4$ mm. From Example 3.6, $v_{full} = 19.1$ mm. Hence from Eq. 3.1, $v_{part} = 19.1 + 0.4 \times (30.4 - 19.1) \times (1 - 0.5) = 21.4$ mm. Decreasing the degree of shear connection from full to 50% therefore increases the long-term deflection by 21.4/19.1 or 12%, which is much less than the increase for the short-term analysis in Example 4.1. Generally speaking, long-term deflections are influenced more by the effects of creep and shrinkage than by partial interaction or partial shear connection.

3.5 Method of construction

3.5.1 *General*

When a composite beam is constructed, it may be *propped* or *unpropped* as shown in Figure 1.3. Propped construction is usually reserved for building construction where the weight of the wet concrete is transferred through the steel component to a number of closely spaced props as in (a). These props are then removed when the concrete has set, and the resulting behaviour is determined by analysing the resulting composite beam. Unpropped construction is usual for bridges, where the steel component alone is required to support its own self weight as well as that of the wet concrete as in (b). Composite action is not achieved from the outset, and the deformations must be calculated initially from the flexural rigidity $E_s I_s$ of the steel component alone. The method of construction has significant ramifications on the flexural stresses, deformations and the forces on the shear connection, and this will be illustrated below by use of examples.

3.5.2 *Flexural stresses*
3.5.2.1 **Example 3.9** *Calculation of flexural stresses in propped beams*

Consider again the T-beam of Example 3.1 which is subjected to positive bending. The uniformly distributed load of 45 kN/m is composed of a long-term dead load w of 15 kN/m and a short-term live load of 30 kN/m. If full interaction is assumed, the stress distributions for long-term loading are shown in Figure 3.7(b) (with $\phi = 3$) and in (c) for short-term loading. The stresses in the transformed section can be calculated from $\sigma_{nc} = My/I_{nc}$ for short-term loading or from $\sigma_{ne} = My/I_{ne}$ for long-term loading. The stresses at midspan calculated using I_{ne} are shown in (b) and those calculated using I_{nc} are shown in (c). In the long-term, $(M_a)_l = 15 \times 6^2/8 = 67.5$ kNm, so that at the top of the transformed section $\sigma_{ne} = 67.5 \times 10^6 \times 171.1/7964 \times 10^6 = 1.5$ N/mm² (Example 3.5) and at the bottom $\sigma_{ne} = -67.5 \times 10^6 \times (454 - 171.1)/7964 \times 10^6 = -2.4$ N/mm² (compressive stresses positive). Similarly in the short-term under a moment $(M_a)_s = 30 \times 6^2 / 8 = 135$ kNm, σ_{nc} (top) = $135 \times 10^6 \times 106.2/2905 \times 10^6 = 4.9$ N/mm (Example 3.1) and σ_{nc} (bottom) = $-135 \times 10^6 \times (454 - 106.2)/2905 \times 10^6 = -16.2$ N/mm². These stresses may be superimposed, and then transformed back to their original constituents according the modular ratios n or n_e as shown in (d). The maximum flexural tensile stress in the steel is thus $7.0 \times 16.2 + 28 \times 2.4 = 181$ N/m².

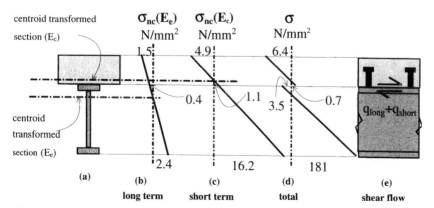

Figure 3.7 Propped construction

3.5.2.2 **Example 3.10** *Calculation of flexural stresses in unpropped beams*

The same beam analysed in Example 3.9 for propped construction is now analysed for unpropped construction. In this typical analysis, it has been assumed that all long-term load is resisted by the steel component, as shown in Figure 3.8(b), and that all short-term load is resisted by the composite beam, as in (c). The long-term stresses are calculated from $\sigma = My/I_s$ in the steel component alone, while those in the composite section due to the short-term loading are calculated from $\sigma_{nc} = My/I_{nc}$. Because the stresses were transformed according to the short-term modular ratio n, they are superposed and transformed according to n, as shown in (d). The short-term stresses are as in Example 3.9, while those in the steel are $\pm 67.5 \times 10^6 \times (324/2)/115.1 \times 10^6 = \pm 95$ N/mm². The maximum flexural tensile stress in the steel is thus $95 + 7.0 \times 16.2 = 208$ N/mm² which is 115% of the maximum flexural stress in propped construction.

3.5.2.3 **Example 3.11** *Deflections in propped and unpropped construction*

We will now consider the midspan deflections for the previous two analyses of propped and unpropped construction. For propped construction, the deflection due to long-term loads (based on I_{ne}) is $(5/384) \times 15 \times 6000^4 / 7.15 \times 10^3 \times 7964 \times 10^6 = 4.5$ mm, and that due to short-term loads (based on I_{nc}) is similarly 6.1 mm, giving a total deflection of 10.6 mm.

In unpropped construction, the deflection due to long-term loading is based on the flexural stiffness $E_s I_s$ of the steel component, and is 11.0 mm. The deflection due to short-term loading is the same as that in propped construction, viz. 6.1 mm, giving a total deflection in unpropped construction of 17.1 mm. It can be seen that the deflections in unpropped construction are considerably higher than those in propped construction.

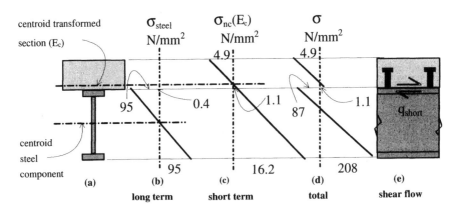

Figure 3.8 Unpropped construction

3.6 Shear flow on connectors

3.6.1 *General*

Assuming linear elastic full interaction analysis, the shear flow on the connectors may be determined from

$$q = \frac{VQ}{I} = \frac{VA_c \, \bar{y}}{I_{nc}} \tag{3.3}$$

where $Q = A_c \bar{y}$ and where $\bar{y}$ is the distance between the centroid of the concrete component and the centroid of the transformed composite beam, that is the composite beam transformed to concrete. In propped construction, the shear flows due to the short and long-term loads have to be determined separately from the short and long-term transformed sections, and the shear flows added as in Figure 3.7(e). On the other hand, in unpropped construction, only the composite behaviour induces shear flows in the connectors, so this will be due to the short-term loading as shown in Figure 3.8(e).

Note that it is more usual to determine the distribution of the shear connectors in composite beams in buildings from the results of a rigid plastic analysis, as described in Chapter 4. However, the following example that uses a beam in a building will demonstrate the use of linear elastic analysis in determining the distribution of shear connectors, as this procedure is generally used in the design of the shear connectors in composite bridge beams.

3.6.1.1 **Example 3.12** *Shear flow on connectors*

For propped construction and with the cross-section considered in the previous examples, the maximum shear for the long-term loading is $V = 45$ kN and $\bar{y} = 171.1 - 130/2 = 106.1$ mm. Hence $q_{long} = 45,000 \times 1500 \times 130 \times 106.1/7964 \times 10^6 = 117$ N/mm and similarly the maximum shear flow for short-term loading $q_{short} = 237$ N/mm producing a total shear flow of 354 N/mm. On the other hand, for unpropped construction only the short-term loads exert forces on the shear connectors, so that $q_{short} = 237$ N/mm as before. Let us assume that we will be designing the shear connectors for propped construction, that is for a shear flow of 354 N/mm based on linear elastic analysis.

It will be assumed that the characteristic strength of a dowel connector is $D_{max} = 60$ kN. The spacing required at the supports is thus $60,000/354 = 170$ mm, and that at midspan is infinite as shown in Figure 3.9. The connectors can be placed in blocks as shown in the figure.

If we are willing to accept an understress of $x = 25\%$ in Figure 3.9, then the length of blocks adjacent to the support is 750 mm where the connectors will have a spacing of 170 mm. If we accept an overstress of 10%, then the length of the next block is $750 + 300 = 1050$ mm where the connectors will have a spacing of 260 mm. The procedure outlined in Figure 3.9 tends to be more conservative at the supports than

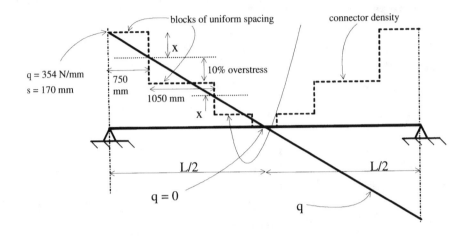

Figure 3.9 Distribution of connectors based on elastic analysis

between the supports, and this is desirable as the shear connectors at the support regions are more prone to failure.

3.7 References

1. Hall, A.S. and Kabaila, A.P. (1986). Basic Concepts of Structural Analysis. GreenwichSoft, Sydney.

2. Bradford, M.A. and Gilbert, R.I. (1992). 'Composite beams with partial interaction under sustained loads'. Journal of Structural Engineering, ASCE, Vol. 118, No. 7, 1871-1883.

3. Johnson, R.P. and May, I.M. (1975). 'Partial interaction design of composite beams'. The Structural Engineer, Vol. 53, 305-311.

4 Rigid plastic analysis of simply supported beams

4.1 Introduction

The maximum possible flexural strength or moment capacity of a simply supported composite beam can be derived from rigid plastic analyses in which it is assumed that all the materials are fully yielded and have unlimited ductility. In order to obtain this upper bound to the flexural strength, it is necessary to ensure that the following modes of failure do not occur prematurely: local buckling of the rectangular elements of the steel component which are dealt with in Chapter 2; lateral-distortional buckling of the steel component as described in Chapter 13; fracture of the shear connectors because of their limited slip capacity as described in Chapter 5; and failure of the concrete component of the composite member due to the concentrated dowel loads imposed on it by the shear connectors as covered in Chapters 6, 10 and 11.

The basic procedures for determining the rigid plastic flexural capacities at a design section of a composite beam are first illustrated in Section 4.2 for standard composite beams in which neither the steel nor the concrete components are encased by the other as shown in Figure 1.1. Non-standard composite beams, in which one component is encased by the other, are analysed in Section 4.3. The distribution of the flexural forces and the terminology used in describing them are explained in qualitative terms in Section 1.4.2 and it is suggested that the reader glance through this before proceeding.

4.2 Rigid plastic flexural capacity of standard composite beams

4.2.1 *Equilibrium of forces at a design section*

It is worth noting that rigid plastic analysis techniques are based purely on equilibrium of forces because all the materials are assumed to be fully yielded and have unlimited ductility. Furthermore, the flexural forces in a composite beam and the positions of the neutral axes depend on the relative strengths of the three components of a composite beam which are the concrete slab, the steel beam and the shear connection.

Consider the simply supported composite beam in Figure 4.1, and let us assume that we are trying to determine the flexural capacity at the design section A-A in (b). The composite beam can be considered to consist of three distinct components which are the concrete and steel components in (a) and the shear connection component in (b). The strengths P of the three components are shown in (c). The compressive strength of the concrete component $P_c = A_c 0.85 f_c$ where A_c is the cross-sectional area of the concrete slab, f_c is the compressive cylinder strength of the concrete

53

which can be taken as 85% of the cube strength f_{cu}, $0.85f_c = f_{cy}$ is the compressive 'yield' strength of the concrete which can also be taken as $0.72f_{cu}$, and where the tensile strength of the concrete material and hence the tensile strength of the concrete component is taken as zero. The tensile and compressive strength of the steel component $P_s = A_s f_y$, where A_s is the cross-sectional area of the steel beam and f_y is the yield strength of the steel. Furthermore, the shear strength of the shear connection is $P_{sh} = Q_{sh} L_{ss}$ where Q_{sh} is the shear flow strength of the shear connection, that is the strength of the shear connection per unit length of beam, and L_{ss} is the length of the shear span between the design section and the support. It is worth noting that while the strengths of the steel and concrete components (P_s and P_c) are unchanged throughout the length of the beam, the strength of the shear connection component (P_{sh}) varies throughout the length of the beam as it depends on the distance from the design section to the support (L_{ss}).

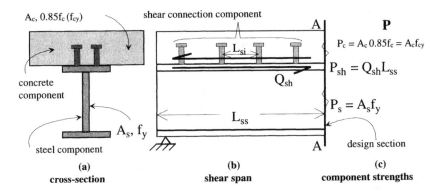

Figure 4.1 Strengths of the components of a composite beam

The forces F acting along the longitudinal axis of a composite beam are shown in Figures. 4.2(a) and (b). As was explained in Section 1.4, the maximum axial force that can act on the concrete component is the weaker of the strengths of the concrete component P_c and of the shear connection component P_{sh}. Similarly, the maximum axial force in the steel component is the weaker of P_{sh} and P_s. Furthermore, the shear forces across the steel/concrete interface in (a) must be equal in magnitude to the axial force and hence equal to the weaker of the forces in the steel and concrete components. Therefore, the component forces C_w in (b) must be the weakest of the strengths of the three components P_c, P_s and P_{sh}, that is the resultant force in the three components of a composite beam C_w is equal to the strength of the weakest of the three components of the composite beam.

Examples of the three possible component force distributions and hence stress distributions are shown in Figures. 4.2(c) to (e). When the steel component is the weakest of the three components, then $C_w = P_s$ in (b), the steel component is fully yielded as in (c) and there is one neutral axis that lies in the concrete component as shown. When the concrete component is the weakest, then

$C_w = P_c$, the concrete is fully 'yielded' at f_{cy} and the one neutral axis now lies in the steel component as in (d). However, when the shear connection component is the weakest, then $C_w = P_{sh}$ and neither the concrete nor the steel components are fully yielded in one direction as shown in (e), that is neither component is fully yielded in either compression or tension, so that there are now two neutral axes. As described in Section 1.4.2, the cases shown in (c) and (d) are often referred to as those of full-shear-connection whereas that shown in (e) is referred to as one of partial-shear-connection.

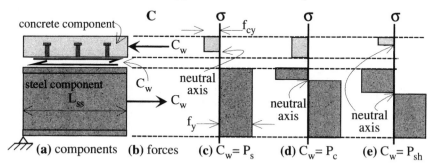

Figure 4.2 Resultant axial forces in components

4.2.2 *Steel component weakest*
4.2.2.1   General

The composite beam shown in Figure 4.3 will be analysed in the following examples to illustrate different aspects of composite construction. The beam has a span of $L_c = 10$ m as shown in (b) and it will be assumed that the composite beams have a lateral spacing of $T_{adj} = 5$ m. Hence, the effective breadth of the concrete component either side of the steel component is $b_{eff} = 1750$ mm as shown in (a); this breadth is based on Ansourian's approach[1] (Eq. 2.8). However, it is worth noting that the Eurocode approach[2] gives an effective breadth of 1250 mm (Eq. 2.3) which is 29% less. As these analyses are based on the steel component being the weakest, it will be assumed that the strength of the shear connection on each side of the design section in Figure 4.3(b), along the length of the shear span $(L_{ss})_l$ and along $(L_{ss})_r$, is equal to or greater than the strength of the steel component. Hence we are dealing with full-shear-connection. Units of N and mm will be used throughout unless stated.

4.2.2.2   **Example 4.1** *Full-shear-connection analysis*
(a) Rigid plastic analysis of a composite beam

The rigid plastic analysis of the composite beam in Figure 4.3 is summarized in Figure 4.4. As we are applying a full-shear-connection analysis, the first step is to determine the distribution of the component forces C in (d). The strength of each rectangular element in (a) is listed in (b), from which it can be seen that the strength of the concrete component is 9692 kN, the strength of the steel component is 2300 kN, and the strength of the shear connection component is shown as P_{sh} which is assumed to be greater than the weaker of the other two components. As the strength of the steel component (2300 kN) is less than the strength of the concrete component (9692 kN), the component forces are equal to the strength of the steel component (that is the weakest element) as shown in (d).

All units in N and mm unless shown

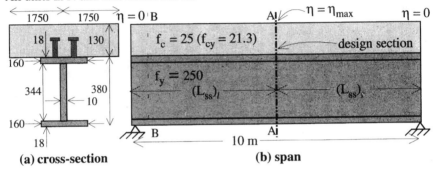

(a) cross-section   (b) span

Figure 4.3 Simply supported standard composite beam

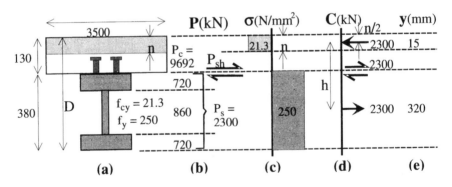

(a)   (b)   (c)   (d)   (e)

Figure 4.4 Example 4.1 Rigid plastic analysis of a composite beam

The neutral axis lies in the stronger component, which in this case is the concrete component, and is shown at a depth n below the top fibre in Figure 4.4 (c). As the steel component is uniformly stressed as shown in (c) and as the steel component is symmetrical, the resultant force in the steel component, shown in (d), acts at the mid-depth of the steel which is 320 mm from the top fibre as in (e). The axial force in the concrete component acts over a depth n and width 3500 mm of the concrete component as shown shaded in (a). Equating this force to the strength of this shaded section, that is 2,300,000 = 21.3 × 3500 × n, gives n = 30.9 mm. Furthermore, the resultant axial force in the concrete component acts at n/2 ≈ 15 mm from the top fibre as shown in (e) where y is the distance of the component force from the top fibre.

The magnitude and positions of all the flexural forces in the composite beam are now known and shown in Figures. 4.4(d) and (e). Taking the moment of the two forces about any convenient axis such as the top fibre or at a position of a resultant component force gives the moment capacity of the composite beam as $M_{fsc} = 2300 \times 0.305 = 702$ kNm. The moment capacity is not sensitive to the effective width of the slab used in the analysis, for example, using the Eurocode effective width (Section 4.2.2.1) which is 29% less than that used in this analysis, reduces the flexural capacity by only 2.0%.

It can also be seen in Figure 4.4(d) that the shear force across the steel/concrete interface is 2300 kN. This is the force in the shear connectors in a shear span such as L_{ss} in Figure 4.2(a). In order to achieve full shear connection, the strength of the shear connectors in each shear span, $(L_{ss})_l$ and $(L_{ss})_r$, in Figure 4.3(b) must be equal to or greater than 2300 kN. Therefore, the *total* strength of the shear connectors in a composite beam must be at least equal to twice the strength of the weaker of the concrete and steel components, which in this example is 4600 kN, in order to achieve full-shear-connection.

(b) Increase in strength due to composite action

A simple and familiar rigid plastic analysis could be used to calculate the moment capacity of the steel beam in Figure 4.3(a), in order to determine the increase in strength due to the composite action. However, the steel beam acting by itself is the composite beam with no shear connection, that is with a zero degree of shear connection. The flexural strength of the steel beam acting by itself will be determined using composite analyses in order to introduce partial-shear-connection analysis techniques; however the concept of partial-shear-connection will be covered in much greater detail in Section 4.2.4.

The partial-shear-connection analysis of the composite beam with zero-shear-connection is illustrated in Figure 4.5. It is worth comparing this analysis with the full-shear-connection analysis shown in Figure 4.4. The strengths of the three components are shown in Figure 4.5(b). The strength of the shear component P_{sh} is the weakest and equal to zero and, therefore, the component forces are all zero as shown in (d). As the concrete component force is zero and as the tensile strength of the concrete is assumed to be zero, the concrete element is unstressed as shown in (c) with the neutral axis at the top fibre. As the steel component force is zero and as the steel component is symmetrical, the neutral axis must lie at mid-depth of the steel component as shown in (c) which gives the forces in (e) and their distances from the steel/concrete interface in (f). Using (e) and (f) to take moments about the interface gives the moment capacity of the steel element as 335 kNm. Therefore, tying the steel beam to the concrete slab using shear connectors has increased the flexural capacity by a factor of 702/335 = 2.1. This substantial increase in strength combined with a similar increase in stiffness (illustrated in Chapter 3) emphasises the enormous gain that can be achieved by making the concrete slab and the steel beam composite.

(c) Approximate approach for initial design

In most composite beams in buildings, the steel component is usually weaker than the concrete component so that the full-shear-connection analysis illustrated in Figure 4.4 will apply most of the time. The moment capacity is equal to the strength of the steel component $(A_s f_y)$ times the lever arm between the forces in the steel and concrete components shown as h in (d). For a composite beam with a symmetrical I-section, the smallest value of h occurs when the strength of the concrete component is equal to the strength of the steel component, in this case the lever arm is equal to half the total depth D of the composite section shown in (a). Therefore and as a first approximation, a lower bound to the full-shear-connection flexural capacity is given by

$$M_{fsc} \leq A_s f_y (D/2) \qquad (4.1)$$

Applying Eq. 4.1 to the composite beam in Figure 4.4 gives a lower bound to the moment capacity of 587 kNm which is 16% less than the upper bound rigid plastic strength of 702 kNm.

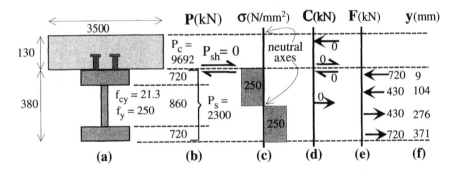

Figure 4.5 Example 4.1 Partial-shear-connection analysis with zero shear connection

4.2.2.3 **Example 4.2** *Efficient forms of composite beams*

The contribution of the top flange of the steel component in Figure 4.4(a) to the moment capacity is small because it is close to the neutral axis. It can therefore be seen that the main purpose of the top flange of the steel component is for the attachment of the shear connectors. An efficient design would be to make the top flange as small as possible or to remove it altogether as shown in Figure 4.6(a), where the shear connectors are welded to the sides of the web of an inverted T-section. Because the cover to the sides of the shear connectors is small in this hybrid composite beam, the concrete element is prone to splitting as described in Chapter 10 where design rules to prevent splitting are given.

The analysis of the hybrid beam is summarized in Figure 4.6; the area of the web encased by the concrete in (a) has been ignored in the analysis. The strengths of the components of the composite beam are listed in (b) where it can be seen that the steel component is the weakest at 1625 kN. Therefore, the steel element is fully yielded in tension as shown in (c) and the strength of the steel component controls the distribution of the component forces C as shown in (d). Because this steel component is not symmetrical, it is much easier to deal with the forces F in the rectangular elements as shown in (e) instead of the resultant component forces in (d). The forces in the rectangular elements of the steel component act at the centroid of the rectangular elements and their distance from the top fibre is shown in (f). The depth of the neutral axis n in (c) is 22 mm and can be derived in the usual way by equating the component force in the concrete element of 1625 kN to the strength of the slab in compression which is $21.3 \times 3500 \times n$.

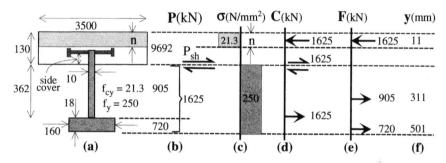

Figure 4.6 Example 4.2 Composite beams with inverted T steel component

By taking moments using Figures. 4.6(e) and (f), the moment capacity of the composite beam without a steel top flange is 624 kNm which can be compared with the capacity of the same composite beam but with the top steel flange in Example 4.1 of 702 kNm. Removal of the top steel flange has reduced the area of steel by 29% but has only reduced the moment capacity by 11%. In order to obtain full-shear-connection, the strength of the shear connection in a shear span must be at least 1625 kN as shown in (d), and hence the strength of the shear connection in the whole beam must be at least 3250 kN.

4.2.2.4 **Example 4.3** *Strengthening composite beams*

The bottom flange in Figure 4.6(a) contributes to most of the moment capacity as it is furthest from the neutral axis and, therefore, an efficient way of increasing the flexural strength of a composite beam is to attach an additional flange as shown in Figure 4.7. This additional flange can have a higher yield strength than that of the I-section making the system even more efficient.

The steps of the analysis are summarized in Figure 4.7. The additional 220×26 mm flange has been chosen to virtually double the strength of the steel component from 2300 kN to 4588 kN. However, the strength of the steel component ($P_s = 2300+2288 = 4588$ kN) is still weaker than that of the concrete component

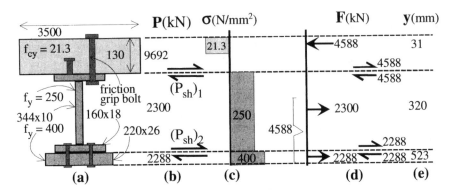

Figure 4.7 Example 4.3 Strengthening composite beams

($P_c = 9692$ kN) so that the steel component still controls the distribution of component forces as shown in (d). From (d) and (e), the moment capacity is now 1790 kNm which is 2.6 times the strength of the original composite beam and 5.3 times the strength of the I-section acting by itself. It can be seen that doubling the strength of the steel component by the addition of a steel flange increases the moment capacity of the composite section by a greater factor (2.6) as the additional flange is placed at its most efficient position.

In order to achieve this increase in the moment capacity, it would be necessary to increase the strength of the shear connectors in a shear span from 2300 kN (as shown in Figure 4.4(d)) to 4588 kN as shown in Figure 4.7(d). This can be achieved with the addition of friction grip bolts as shown in (a). However, if it is impractical to add more shear connectors, then the increase in strength can be derived from partial-shear-connection analyses as in Example 4.7. The additional 220 × 26 mm flange has to be attached to the I-section by bolting or welding as shown in Figure 4.7(a) and the strength of this shear connection per shear span must be at least equal to 2288 kN as shown in (d) in order to achieve full-shear-connection for the additional flange.

4.2.2.5 **Example 4.4** *Composite beams with longitudinal ribs*

The previous examples have dealt with composite beams with solid slabs. An alternative and common form of construction is a composite beam that has a composite profiled slab as its concrete component. The ribs of these composite slabs can be longitudinal to the composite beam as shown in Figure 4.8(a) or they can be transverse to the composite beam as in Figure 4.10. As with the hybrid beam in Figure 4.6(a), composite beams with longitudinal ribs as in Figure 4.8(a) are prone to splitting because of the limited side cover to the shear connectors. It is up to the designer to determine whether the splitting resistance, as determined in Chapter 10, controls the strength of the shear component. In order to design the composite beam in Figure 4.8(a), we will use the effective section in (b) as described in Section 2.2.3, which has the same cross-sectional area of slab as in (a). To further simplify the problem, we will assume that the haunch has vertical sides as shown in (b) as the error is minuscule.

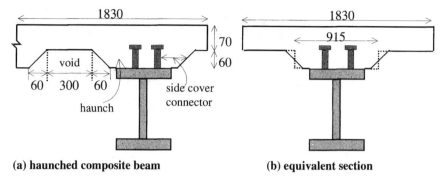

(a) haunched composite beam   **(b) equivalent section**

Figure 4.8 Equivalent section of a composite beam with longitudinal ribs

The composite beam in Figure 4.8(b) has been analysed in Figure 4.9 for different yield strengths of the steel of 250 N/mm² and 400 N/mm². For f_y = 250 N/mm², the strengths of the elements are shown in (b). It can be seen that the strength of the steel component (2300 kN) is less than the strength of the concrete component (2722 + 1166 = 3888 kN). In fact the strength of the steel component is smaller than the strength of the upper element of the concrete component (2722 kN), and therefore the neutral axis lies in this upper element, that is above the rib. Because the neutral axis lies above the rib, the analysis is the same as that of the composite beam with a solid slab in Figure 4.4, except that the width of the slab is now 1830 mm instead of 3500 mm in Figure 4.4. The depth of the neutral axis n = 2,300,000/(1830 × 21.3) = 59 mm and hence the moment capacity is now 2300 × 0.2905 = 668 kNm. It can be seen that virtually halving the width of the slab from 3500 mm to 1830 mm has only reduced the strength from 702kNm to 668 kNm that is by 5%.

The analysis when f_y = 400 N/mm² is shown in Figures. 4.9(c) to (f). The strength of the steel component (3680 kN) is still weaker than the strength of the concrete component (2722 + 1166 = 3888 kN) but stronger than the strength of the upper element of the concrete component. Therefore, the neutral axis now lies in the lower element of the concrete component as shown in (d) and, furthermore, the resultant force in each component is equal to the strength of the steel component (3680 kN) as shown in (e). The distribution of the force in the concrete component of 3680 kN consists of 2722 kN in the upper element with the remainder of 958 kN in the lower element as shown in (e). If n is the depth of concrete in compression in the lower concrete element as shown in (a) and (d), then by equating the element force (958 kN) to the strength of the concrete element in compression (915×n×21.3) gives n = 49.2 mm. The resultant forces and their distance from the top fibre are shown in (e) and (f) from which it can be determined that the moment capacity is 991 kNm.

The analysis procedure described in the previous paragraph was based on the fact that the resultant force in each component is equal to the strength of the weakest component. An alternative way of visualizing the problem is that the compressive force above the neutral axis is equal to the tensile force below the neutral axis. Take for example the beam in Figure 4.9 with f_y = 400 N/mm² and where the strengths of

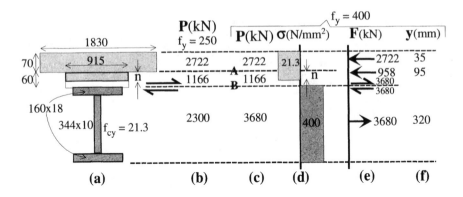

Figure 4.9 Example 4.4 Composite beam with longitudinal ribs

the elements are listed in (c). The element in which the neutral axis occurs can be determined by trial and error. For example if the neutral axis were assumed to be at level A, the resultant force is tensile and equal to $3680 - 2722 = 958$ kN, however, at level B the resultant force is compressive and equal to $3680 - 2722 - 1166 = -208$ kN. Therefore, the neutral axis lies between levels A and B.

4.2.3 *Concrete component weakest*
4.2.3.1   General

Occasionally the concrete component is weaker than the steel component. This can happen in composite L-beams such as that shown in Figure 4.8(a), particularly if the concrete component is a composite slab as the profiled ribs reduce the area of concrete. This is also often the case in unpropped composite bridge beams where deflection is a major design criterion necessitating a large steel element. It will again be assumed in the following analyses that the strength of the shear connection in a shear span is greater than the strength of the concrete component and hence we are still dealing with full-shear-connection analyses.

4.2.3.2   **Example 4.5** *Full-shear-connection analysis of a composite beam with transverse ribs*

(a) Rigid plastic analysis

A composite L-beam that has a composite slab with transverse ribs is shown in Figure 4.10. The flexural strength is governed by the weakest cross-section which occurs between the transverse ribs such as at A-D in (b). Therefore, the section to analyse has a solid slab of depth A-B and a steel element of depth C-D that is separated by the void due to the ribs of depth B-C as shown in Figure 4.11(a).

By inspection of the element strengths in Figure 4.11(b), the neutral axis lies in the top steel flange. If the neutral axis lies at a distance n below the concrete/steel interface as shown in (c), then equating the compressive force above the neutral axis $(2,729,000 + (160 \times n \times 400))$ to the tensile force below $(160(18 - n)400 + 1,376,000$

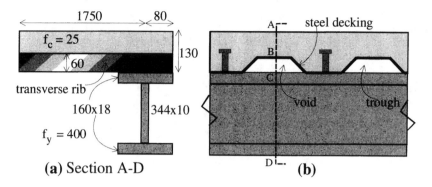

Figure 4.10 Composite L-beam with transverse ribs

+ 1,152,000) gives n = 7.4 mm and, hence, the magnitudes and positions in (d) and (e), from which can be determined the moment capacity of 955 kNm. It can be seen in (d) that the strength of the shear connection per shear span must be 2729 kN.

(b) Equivalent stress block approach

The strength and stress distributions in Figures. 4.11(b) and (c) are shown in Figures. 4.12(a) and (b). A simpler analytical approach is to use the equivalent stress system in (c) in which all of the steel element is yielded in tension at f_y and the part that is in compression has an increased stress of $2f_y$. It can be seen that this equivalent stress distribution has the same resultant stress distribution as in (b).

Consider the equivalent stress system of Figure 4.12(c). The tensile force in the steel component is 3680 kN as shown in (d) and as the resultant force in the steel component is 2729 kN, the compressive force is 951 kN as shown. If the depth of the neutral axis is n as in (c), equating the compressive force (951 kN) to the strength of the steel flange in compression ($160 \times n \times 400 \times 2$) where the yield strength in compression is $2f_y$, gives n =7.4 mm and the position of this compressive force from

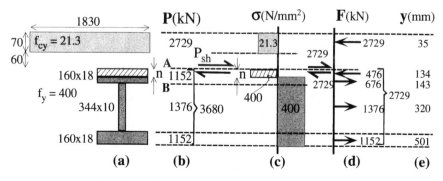

Figure 4.11 Example 4.5 Full-shear-connection analysis

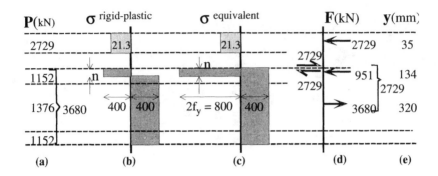

Figure 4.12 Example 4.5 Equivalent stress block approach

the top fibre as 134 mm in (e). It can be seen that deriving the moment capacity from Figures. 4.12(d) and (e) is much simpler than from Figures. 4.11(d) and (e).

4.2.4 *Shear component weakest*
4.2.4.1   General
When the strength of the shear component is the weakest of the three components, then the component forces are equal to the strength of the shear component and this is referred to as partial-shear-connection. Partial-shear-connection can occur in com-posite beams with transverse ribs, as in Figure 4.10(b), because there are a limited number of troughs through which the connectors can be welded and also because the voids either side of the troughs weaken the shear connection as explained in Chapter 5. Furthermore, a composite beam may start with full-shear-connection, but the shear connection may weaken with time due to splitting (Chapter 10) or fatigue (Chapter 15), so that the beam eventually has partial-shear-connection. Or quite simply, the designer may find that the full-shear-connection strength is more than required and hence uses fewer shear connectors to reduce both the strength and the cost.

4.2.4.2   **Example 4.6** *Partial shear connection analysis*
The composite beam in Figure 4.4 was originally designed with full-shear-connection where the strength of the shear connection in a shear span was at least 2300 kN as shown in (d). Let us assume that the traversal of a concentrated load caused splitting and that the post-splitting strength (Chapter 11) is 20% weaker at 1840 kN. The composite beam now has partial-shear-connection and the degree of shear connection as defined in Section 1.4.2.3 and Eq. 1.13 is 80%. The analysis of the composite beam is shown in Figure 4.13 where the strengths of the components are shown in (b). As the shear component is the weakest component, the resultant force in all three components is equal to the strength of the shear component as shown in (c) and, furthermore, there are now two neutral axes as shown in (d). The equivalent stress distribution in (e) will be used in the following analyses.

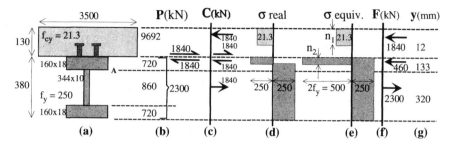

Figure 4.13 Example 4.6 Partial-shear-connection analysis

The concrete component is not fully stressed in compression as shown in Figure 4.13(e). The neutral axis position n_1 can be determined by equating the component force (1840 kN) to the strength of the concrete compression zone ($3500 \times n_1 \times 21.3$) to give $n_1 = 24.7$ mm. The steel component is also not fully stressed in tension. The neutral axis position n_2 can be determined by inspection. Consider the strengths of the steel rectangular elements in (b) and recall that the resultant force in the steel component is 1840 kN as in (c). If the neutral axis is at level A in (b), then the resultant tensile force in the steel component is the tensile force below the neutral axis less the compressive force above, that is $(720 + 860) \times 720 = 860$ kN which is less than the required value of 1840 kN. Therefore the tensile force has to be increased by raising the neutral axis above level A, that is into the top steel flange where the neutral axis is shown at a distance n_2 in (d). The force in the compression zone of the steel element in (e) must equal $2300 - 1840 = 460$ kN and equating this force to the strength of the compression zone ($160 \times n_2 \times 500$) gives $n_2 = 5.8$ mm.

From Figures. 4.13(f) and (g), the moment capacity is 653 kNm. It is worth noting that a 20% reduction in the strength of the shear connection from full shear connection has only reduced the moment capacity from 702 kNm to 653 kNm, that is by 7%. It can be seen that the moment capacity of a composite beam with an initial high degree of shear connection is not sensitive to reductions in the strength of the shear connection, so that composite beams can generally withstand substantial damage to the shear connection with minimal effect on their flexural capacity.

4.2.4.3 **Example 4.7** *Strengthening of composite beams*

The composite beam in Figure 4.3 was strengthened in Figure 4.7 by adding a flange and extra shear connectors. Let us now assume that it is impractical to add more shear connectors and hence the composite beam now has partial-shear-connection, and the degree of shear connection is the strength of the original shear connection divided by that required for full-shear-connection, that is $\eta = 2300/(2300 + 2288) = 50\%$.

The analysis of the beam is shown in Figure 4.14. The elements in which the neutral axes can occur can be determined by inspection. Equating the force in the concrete element (2300 kN) to the strength of the area in compression ($3500 \times n_1 \times 21.3$) gives $n_1 = 30.9$ mm.

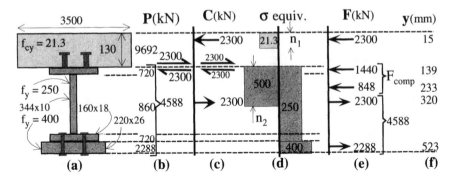

Figure 4.14 Example 4.7 Strengthening a beam with partial-shear-connection

The steel component force is 2300 kN as shown in Figure 4.14(c), the tensile strength of the steel component is 4588 kN as in (b), and the compressive strength of the top steel flange is $2 \times 720 = 1440$ kN as in (e) because the compressive yield strength of the top flange is assumed to be $2f_y = 500$ N/mm^2 as shown in (d). As the steel component force is 2300 kN, from (e) it can be seen that $4588 - F_{comp} = 2300$ where F_{comp} is the compressive force in the steel component with the stress distribution in (d). Therefore $F_{comp} = 2288$ kN. The force F_{comp} is resisted by the top flange and part of the web as shown in (d) hence the compressive force in the web is $2288 - 1440 = 848$ kN which is equal to the strength of the web in compression ($10 \times n_2 \times 500$), and hence $n_2 = 169.6$ mm. The magnitudes of the forces and their positions are now know and shown in (e) and (f) from which the moment capacity is 1500 kNm.

It can be seen from the calculations in Figure 4.14 that, even though additional shear connectors were not added to provide full-shear-connection, the inclusion of the plate has substantially increased the moment capacity from 702 kNm to 1500 kNm. Another way of viewing this situation is that a 50% reduction in the shear connector capacity of the composite beam from full-shear-connection has only reduced the flexural capacity by 16% from 1790 kNm to 1500 kNm. It is worth noting that the strength of the bolt shear connectors shown in Figure 4.14(a) must still remain at 2288 kN per shear span as the force in the additional plate is unchanged. It is also worth noting that composite beams with low degrees of shear connection are prone to premature failure due to fracture of the shear connectors as described in Chapter 5.

4.2.5 *Effect of vertical shear on the flexural capacity*
4.2.5.1 Equivalent flexural yield strength
The vertical shear force in a composite beam is assumed to be resisted entirely by shear stresses in the web of the steel element. These shear stresses τ_w reduce the flexural stress that cause the web to yield from the yield strength f_y to an equivalent yield strength f_{fy} that can be derived from von Mises yield criterion described in Section 1.3.2 as

$$f_{fy} = \sqrt{f_y^2 - 3\tau_w^2}$$
(4.2)

where the shear stress τ_w is assumed to be uniformly distributed over the web.

4.2.5.2 **Example 4.8** *Reduction in flexural capacity due to vertical shear forces*

Let us determine the reduction in the flexural capacity due to vertical shear forces for the composite beam in Figure 4.3 when it is subjected to a uniformly distributed load and when the composite beam has a uniform distribution of shear connectors. From Section 4.2.2.2, the composite beam has a full-shear-connection moment capacity of 702 kNm. Therefore, the beam can support a uniformly distributed load of 56.2 kN/m over its span of 10 m which gives reactions at the supports of 280.8 kN.

As there is a uniformly distributed load, the vertical shear force at mid-span is zero and hence the full-shear-connection moment capacity of 702 kN is not reduced at this section. At the supports, the vertical shear force is at its greatest but the applied moment is zero, so the effect of vertical shear is irrelevant unless the shear stress exceeds $f_y/\sqrt{3}$. Instead, let us consider the effect of the vertical shear on the beam at the quarter-span where the vertical shear force is 140.4 kN. At the quarter-span, the degree of shear connection is 50% as there is a uniform distribution of shear connectors, that is the strength of the shear connection between the quarter-span and the support is half the strength between mid-span and the support.

The analysis at the quarter-span is shown in Figure 4.15. The average shear stress in the web is $\tau_w = 140,400/(344 \times 10) = 40.8$ N/mm² and hence from Eq. 4.2 $f_{fy} = 239.8$ N/mm² which is only a slight reduction from the yield strength of 250 N/mm². The strength of the elements are shown in Figure 4.15(b) where the strength of the web of 825 kN is based on f_{fy}. The analysis follows the usual procedure in (c) to (e) which gives a moment capacity of 562.8 kNm. If the effect of the vertical force is ignored, then the strength of the web in (b) is now 860 kN instead of 825 kN and the moment capacity is 569.3 kNm. Hence, the vertical shear force has only reduced the moment capacity by 1.1% which is irrelevant and generally ignored in practice. However, it is worth noting that the vertical shear force has a much greater effect in negative regions where the position of the maximum shear and maximum moment coincide (Chapter 12), and also in the vicinity of service ducts (Chapter 9) where a mechanism is required to transfer the shear forces across the duct.

4.3 Rigid plastic flexural capacity of encased composite beams

4.3.1 *General*

The full-shear-connection analysis of an encased composite beam differs from the analysis of a standard composite beam because one component is now encased by the other and, therefore, the neutral axis must now lie in both components.

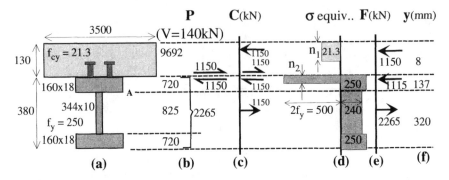

Figure 4.15 Example 4.8 Effect of vertical shear forces at quarter-span

4.3.1.1   **Example 4.9** *Infilled box section*
(a) Full shear connection analysis

The full-shear-connection analysis of an infilled box section is shown in Figure 4.16. The strengths of the rectangular elements are shown in (b), the stress distribution in the concrete component in (c), and the equivalent stress distribution in the steel component in (d). The neutral axis position in (c) and (d) can be determined by bearing in mind that the total force above the neutral axis, that is in the steel and concrete elements, is equal to the total force below. Hence by inspection, the neutral axis lies below the inside of the top flange at the distance n as shown. Using the equivalent stress system in (d), the tensile force of 924 kN in (e) is the strength of the steel component, the 'strength' of the steel flange in compression is 360 kN and, therefore, the remaining steel and concrete in compression must resist a force of $924 - 360 = 564$ kN. Equating this force to the sum of the 'strength' of the steel in compression $(12 \times n \times 500)$ and the strength of the concrete in compression $(29.8 \times n \times 108)$ gives n = 61.2 mm.

The magnitudes and positions of all the forces in Figures. 4.16(e) and (f) are now known, from which the moment capacity is 70.5 kNm. This is only slightly more than the moment capacity of the steel component acting by itself of 61.4 kNm. However,

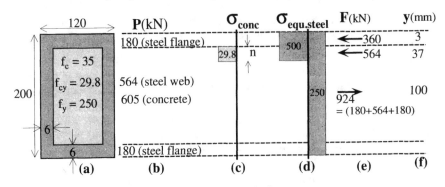

Figure 4.16 Example 4.9 Full-shear-connection analysis

it is worth noting that the addition of the concrete substantially increases the local buckling strength of the steel elements as now they can only buckle away from the concrete component and, furthermore, the addition of the concrete substantially increases the flexural ductility of the structure making it ideal for earthquake zones.

As with the standard composite beam in Section 4.2, the resultant force in the concrete component in Figure 4.16(c) is equal to the resultant force in the steel component in (d) which is equal to the resultant force in the shear connection component in a shear span. The component force, and hence the shear connector force in a shear span, can be determined from the resultant force in either (c) or (d) and is equal to 197 kN. This shear connector force acts uniformly around the perimeter of the interface between the steel and concrete components (because the slip-strain throughout the depth of the beam is constant[3]) of length 592 mm, and on a area extending the full length of the shear span, which we will assume to be 2 m. Therefore, the bond strength required for full-shear-connection is $(197000/(592 \times 2000)) = 0.17$ N/mm².

(b) Partial shear connection analysis

Let us assume that, in the previous example in Section 4.3.1.1(a) above, the strength of the shear connector component is less than the 197 kN required for full shear connection and is equal to 100 kN. We are therefore dealing with a composite beam with partial-shear-connection and, therefore, the resultant force in both the steel component and the concrete component is 100 kN. The partial-shear-connection analysis is shown in Figure 4.17.

The resultant force in each component is now 100 kN as shown in Figure 4.17(c). Because there is partial-shear-connection, there are two neutral axes as shown in (d) for the concrete component and in (f) for the steel component, and their positions can be determined by considering each component separately. The distribution of forces in the steel component is shown in (g). For the resultant force to be 100 kN, the neutral axis must lie below the top flange as shown in (f) and the 'compressive' force below the top flange F_{comp} can be derived from $924 - 360 - F_{comp} = 100$ kN and hence $F_{comp} = 464$

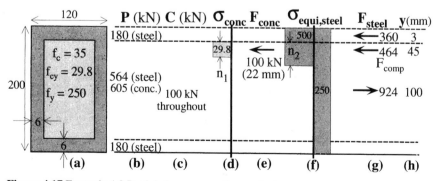

Figure 4.17 Example 4.9 Partial-shear-connection analysis

kN. Equating this force to the 'compressive' strength of the steel component below the top flange ($12 \times n_2 \times 500$) gives $n_2 = 77.3$ mm. The positions of all the forces from the top fibre are shown in brackets in (h). Using a similar procedure for the concrete component gives $n_1 = 31.1$ mm and the position of the force in brackets in (e). The moment capacity can be determined from (e) and (h) and is equal to 68.2 kNm.

4.4 Variation of flexural capacity along the length of the beam

4.4.1 *General*

In all of the previous sections of this chapter, we dealt with the flexural capacity at a design position, such as at section A-A in Figure 4.3(b). This design section is usually chosen at the position of the maximum applied moment. The strength of the shear connector component is the strength of the shear connectors in a shear span, and it is necessary to ensure that this strength is the same in the shear spans on either side of the design position. The maximum degree of shear connection occurs at this design position which will be referred to as η_{max}.

The shear connectors have to be carefully distributed within each shear span in Figure 4.3(b) in order to ensure that the variation of the flexural strength along the length of the beam is never exceeded by the distribution of the applied moment along the length of the beam. Furthermore, it is also necessary to choose a distribution of shear connectors that ensures that the connectors do not fracture prematurely due to excessive slip as described in Chapter 5.

4.4.2 *Uniformly distributed shear connection*

4.4.2.1 General

Most composite beams in buildings are designed to resist an applied load that is uniformly distributed along the length of the beam. It is normal practice to design these beams with a uniform distribution of shear connection, that is the shear flow strength of the shear connectors is constant throughout the length of the beam.

4.4.2.2 **Example 4.10** *Variation in the moment capacity*

(a) Equilibrium approach

A half-span of the composite beam in Figure 4.3 is shown in Figure 4.18(a). The beam is subjected to a uniformly distributed applied load and the shear connectors are also uniformly distributed as shown. The beam was analysed in Figure 4.4 for full-shear connection where it was shown that the moment capacity is 702 kNm. It was also shown that the strength of the shear connection in a shear span required for full-shear-connection, that is $\eta_{max} = 1$, is 2300 kN and, therefore, the beam requires a shear-flow-strength of $Q_{sh} = 2300/5 = 460$ kN/m.

As the beam in Figure 4.18(a) is subjected to a uniformly distributed load, the maximum applied moment occurs at mid-span at section A-A where the degree of shear connection $\eta = \eta_{max} = 1$. Consider section B-B at a quarter span. The strength of the shear connection in the shear span to the left of this design point is $Q_{sh}(L_{ss})_l = 460 \times 2.5 = 1150$ kN, and the strength of the shear connection in the shear span to the right is $Q_{sh}(L_{ss})_r = 460 \times 7.5 = 3450$ kN. The maximum thrust that the connectors can apply at this design point is the weaker of these two strengths, that is 1150 kN, and, therefore, the degree of shear connection at the quarter-span is $1150/2300 = 0.5$. Similarly at section C-C which is adjacent to the support, the strength of the shear connectors to the left tends to zero so that $\eta \rightarrow 0$. It can be seen that the degree of shear connection varies along the half span from $\eta = 0$ at the supports to $\eta = \eta_{max} = 1$ at mid-span and because there is a uniform distribution of shear connectors, the variation in the degree of shear connection along the shear span is linear as shown in (c).

In the previous worked examples, the moment capacities at degrees of shear connection of 0, 0.5, 0.8 and 1 were calculated and these have been plotted in Figure 4.18(b) as the 'moment capacity' curve. It can be seen that the moment capacity varies from the strength of the steel component acting by itself ($M_s = 335$ kNm from Example 4.1(b)) at the supports, to the full-shear-connection moment capacity ($M_{fsc} = 702$ kNm from Example 4.1(a)) at mid-span. Also plotted in (b) is the 'applied moment' distribution which varies from zero at the supports to that of the moment capacity at mid-span. In this example, the applied moment exceeds the moment capacity in region D-E and hence the beam will be slightly weaker than anticipated. This can be corrected by adding more connectors or moving the connectors towards the supports but this slight difference is usually ignored in practice.

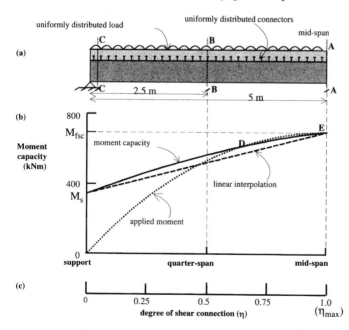

Figure 4.18 Example 4.10 Uniform distribution of shear connectors

(b) Linear interpolation approach

The design procedure can be simplified by assuming a linear variation in the moment capacity from $M_s = 335$ kNm to $M_{fsc} = 702$ kNm as shown by the 'linear interpolation' line in Figure 4.18(b). For example, let us assume that we only require a moment capacity of 569 kNm, hence, from the 'linear interpolation' line in (b) we require a degree of shear connection of $\eta_{max} = (569 - 335)/(702 - 335) = 0.64$. Therefore to achieve a strength of 569 kNm, the beam can be designed with partial-shear-connection and with only 64% of the shear connectors required for full-shear-connection. The linear interpolation approach is conservative. If the more accurate equilibrium approach had been applied, then from Example 4.8, the degree of shear connection would be 50% instead of 64%.

4.4.3   *Non-uniformly distributed shear connection*
4.4.3.1   General

When beams are subjected to concentrated loads or variable distributed loads, then it may become unsafe to use a uniform distribution of shear connectors, as the applied moment may substantially exceed the moment capacity in regions along the shear span or fracture of the shear connectors due to excessive slip may occur. The designer could guess a configuration of connectors and then check for strength at various design positions using the procedure described in Section 4.4.2 and also check for fracture using Chapter 5, however, this may be impractical. Instead, the designer may use guidelines or rules of thumb. One such guideline is to concentrate the connectors according to the distribution of the vertical shear force V as the longitudinal linear elastic shear flow force q is proportional to V, that is $q = VQ/I = VA\bar{y} /I$. It should be remembered that this is only a guideline as we are dealing with rigid plastic theory and not linear elastic theory from which the VAy equation is derived.

4.4.3.2   **Example 4.11** *Distribution of shear connectors*

The beam in Figure 4.3 has a shear connector strength per shear span of 2300 kN and is subjected to the applied loads in Figure 4.19(a). The vertical shear force distribution is shown in (b) from which it can be seen that the position of maximum moment (zero vertical shear force) occurs at mid-span, and hence the shear spans are of equal length of 5m and each requires a strength of shear connection of 2300 kN.

The areas of the shear force diagram between the concentrated loads in Figure 4.19(a) is shown in brackets in (b). It is normal practice to distribute the connectors according to these areas. Therefore in the span A-B, the required strength of the shear connectors is $2300 \times (542/(542 + 160)) = 1777$ kN as shown in (c), and similarly the strength of the shear connection in B-C in (b) is $2300 \times (160/(542+160)) = 523$ kN. As C-D in (b) is a shear span, the strength of the shear connectors is 2300 kN as shown in (c). Therefore, the mean shear flow strength required in span A-B is $1777/2.5 = 711$ kN/m as shown in (d) and similarly that required in B-C is 209 kN/m, and that required in C-D is 460 kN/m. It is worth noting that even though

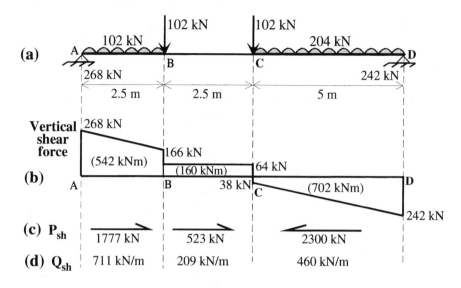

Figure 4.19 Example 4.11 Distribution of shear connectors

there is a linear variation in the linear elastic shear flow force in regions A-B and C-D (as shown by the linear variation in V in (b)), the shear flow strengths are kept constant in these regions as shown in (d), which is in line with the procedure described in Section 4.4.2 where the shear flow strength was kept constant in a beam with a uniformly distributed load.

4.5 References

1. Ansourian, P. (1975). 'An application of the method of finite elements to the analysis of composite floor systems'. Proceedings of the Institution of Civil Engineers, London, Part 2, Vol. 59, 699-726.
2. Eurocode 4 (1994). Part 1: Design of Composite Steel and Concrete Structures. DDENV 1994-1-1: 1994. Draft for development.
3. Oehlers, D.J. and Bradford, M.A. (1995). Composite Steel and Concrete Structural Members: Fundamental Behaviour. Pergamon Press, Oxford.

5 Mechanical shear connectors

5.1 Introduction

The linear elastic shear flow forces in composite beams were derived in Chapter 3 and the rigid plastic shear flow forces in Chapter 4. These shear flow forces have to be resisted by the shear flow strengths of the mechanical shear connectors that are used to tie the concrete component to the steel component. There is an enormous variety of mechanical shear connectors as shown in Figure 1.2 so their properties are always determined experimentally in simple push-tests where the shear load is applied directly to the shear connection[1]. Furthermore, all of these mechanical shear connectors resist the shear flow forces by acting as steel dowels embedded in a concrete medium as shown in Figure 5.1, they all require slip between the concrete component and the steel component to resist these shear forces which are also shown, and they all have to be able to prevent the concrete component from separating from the steel component which is the purpose of the head of the stud.

This chapter will only deal with stud shear connectors, which are unthreaded bolts that are welded to the steel component and then encased in concrete as in Figure 5.1, as these are the most common form of shear connection. The diameter of the shank d_{sh} varies from about 13 mm to 22 mm, with 19 mm being a common size for use in composite beams in buildings. The head of the stud is about $1.5d_{sh}$ wide and $0.5d_{sh}$ deep, and the weld collar is about $1.3d_{sh}$ wide and varies in height h_{wc} from zero to about $0.4d_{sh}$. The height of the stud h_{st} is usually greater than $4d_{sh}$ with a common size being $5d_{sh}$.

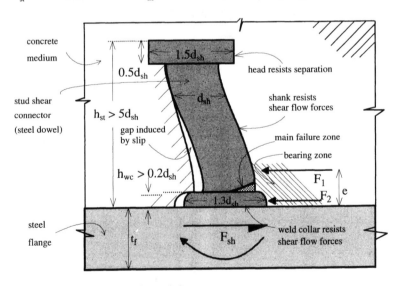

Figure 5.1 Dowel action of a stud shear connector

A typical experimentally determined shear-load/slip characteristic of a stud shear connection is shown in Figure 5.2. The initial response O-B is reasonably linear up to a slip s at S_p of about $0.1d_{sh}$ at which point plasticity commences. There is then a plastic plateau B-C at the maximum dowel strength D_{max}, until the stud fractures due to excessive slip at S_{ult} which occurs at about $0.3d_{sh}$. The response is not unlike the stress/strain relationship of many steels as shown in Figure 1.8. However, unlike steel where the strain at fracture is about 100 times the yield strain, the slip at fracture S_{ult} is only about 3 times the slip at which plasticity commences S_p. Hence, a major concern in composite beam design is to ensure that the connectors do not fracture prematurely due to excessive slip.

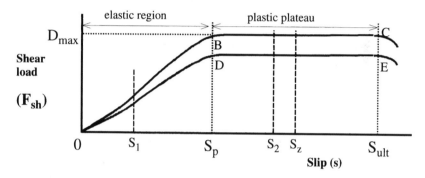

Figure 5.2 Load/slip characteristics of a stud shear connector

The analyses presented in this chapter are based on the maximum dowel strengths D_{max} of stud shear connectors that can be achieved when all the other failure modes that are described in Chapters 6 and 10 are designed against. Detailing rules are first given to ensure that this dowel strength can be achieved, this is followed by methods for determining the dowel strengths and the distribution of connectors in a beam and, finally, procedures are described to ensure that these connectors do not fracture prematurely due to excessive slip.

5.2 Local detailing rules

5.2.1 *General*

The transfer of the longitudinal shear by the dowel action of the stud shear connection exerts very high stresses onto the concrete surrounding the stud and to the steel plate to which the stud is welded. For example, the concrete in the bearing zone shown in Figure 5.1 has to be restrained triaxially to resist the bearing stresses that have to exceed 10 times the cylinder strength of the concrete f_c in order for the dowel strength to be achieved. The distribution of these stresses, that are local to the dowel, is extremely complex, so we have to resort to empirically derived detailing guidelines to ensure that premature failure does not occur due to these local stresses.

5.2.2 *Shape of stud shear connections*

The maximum dowel strength D_{max} varies from about 50 kN to 250 kN, depending on the diameter of the shank and the material strengths of the concrete f_c and the stud f_u. The dowel strength also depends on the height of the stud, as short studs tend to pull out of the concrete. Research[1] suggests that the maximum dowel strength is achieved when the height of the stud $h_{st} \geq 5d_{sh}$ as shown in Figure 5.1. Research has also shown[1] that the dowel strength depends on the height of the weld collar h_{wc} as the weld collar has a much larger cross-sectional area than the shank of the stud and, hence, can resist a substantial amount of the longitudinal thrust. The analyses in this chapter will be limited to groups of stud shear connectors where the mean height of the weld collars $h_{wc} \geq 0.2d_{sh}$ and where $h_{st} \geq 5d_{sh}$.

The dowel action of a stud shear connector imposes high stress concentrations on the steel flange of thickness t_f in Figure 5.1. In order to prevent the flange tearing, it is recommended that the flange thickness $t_f \geq 0.4d_{sh}$ and that the distance between the edge of the connector and the edge of the flange $L_e \geq 1.3d_{sh}$ as shown in Figure 5.3(a).

5.2.3 *Spacing of stud shear connectors*

In the dowel mechanism for transferring the shear flow forces, a small volume of concrete adjacent to both the weld collar and the shank in the bearing zone in Figure 5.1 is crushed. In order to ensure that this failure zone does not affect adjacent connectors, the longitudinal spacing of the connectors $L_L \geq 5d_{sh}$ as shown in Figure 5.3(b) and the transverse spacing $L_T \geq 4d_{sh}$. Furthermore, in order to limit the separation between the steel and concrete components in the region between the connectors, so that we can assume that the curvatures are the same in both components, the longitudinal spacing $L_L \geq 6h_c$ as in (b) where h_c is the depth of the slab as in (a). It is also suggested that, wherever possible, the connectors are staggered over the width and length of the flange as shown in (c), in order to prevent a longitudinal crack forming in the concrete component due to the splitting action of the shear connectors[1], and in order to prevent shear failure of the concrete component as described in Chapter 6. The alternative of

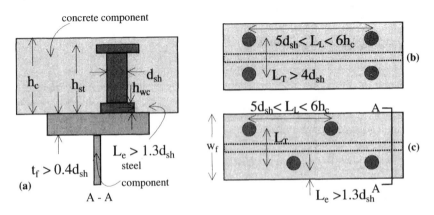

Figure 5.3 Detailing rules for stud shear connections

a single line of stud shear connectors may be aesthetically pleasing to an engineer but can cause the connectors to slice through the slab prematurely.

5.3 Dowel resistance to the shear flow forces

5.3.1 *General*

The shear flow forces are usually derived from rigid plastic analyses (Chapter 4) for composite beams in buildings and from linear elastic analyses (Chapter 3) for composite beams in bridges. These shear flow forces are resisted by the shear flow strengths of the stud shear connectors. These strengths are derived empirically from push tests[1] in which the shear connection is loaded directly, in comparison with composite beams where the shear connection is loaded indirectly through the flexural action of the beam. In the following section, several procedures will be described for determining the shear flow strengths and these will be used to determine the distribution of connectors in composite beams in buildings as well as in bridges.

5.3.2 *Mean strength in push-tests approach*

5.3.2.1 Mean dowel strength of stud shear connections in push-tests

The mean strength of stud shear connections in push-tests[2] is given by the following equation.

$$\left(D_{max}\right)_{push} = 0.50 A_{sh} \sqrt{f_c E_c} \tag{5.1a}$$

when

$$f_u \geq 486 \ N \ / \ mm^2 \tag{5.1b}$$

where A_{sh} is the cross-sectional area of the shank of the stud and E_c is the short term Young's modulus for the concrete surrounding the stud, and in which the ultimate tensile strength of the stud material in these tests was $f_u = 486 \ N/mm^2$. The most important parameter in Eq. 5.1 is the cross-sectional area of the shank A_{sh} at the level of the main failure zone shown in Figure 5.1 where the critical flexural and shear stresses are resisted. The concrete parameter $\sqrt{f_c E_c}$ in Eq. 5.1 affects the position e of the shear force F_1 in Figure 5.1 and hence the flexural stresses in the shank. For normal density concrete, the material stiffness of the concrete E_c can be derived from Section 1.3.5.1 where

$$E_c = 5050\sqrt{f_c} \tag{5.2}$$

when E_c and f_c are measured in N/mm².

5.3.2.2 **Example 5.1** *Mean dowel strength in push-tests approach*
(a) Composite beam

Let us determine the number and distribution of the stud shear connectors for the simply supported composite beam of span 10 m in Figure 4.3, which has full-shear-connection and when the beam is subjected to a uniformly distributed load. The strength of the concrete $f_c = 25$ N/mm², the depth of the slab $h_c = 130$ mm, and the thickness and width of the flange to which the studs are welded are $t_f = 18$ mm and $w_f = 160$ mm. As the beam is subjected to a uniformly distributed load, the maximum moment occurs at mid-span and, hence, the shear span is half the length of the beam, that is $L_{ss} = 5$ m, and, furthermore, the connectors need to be distributed uniformly along a shear span.

The rigid plastic full-shear-connection analysis for this composite beam is given in Example 4.1 in Section 4.2.2.2 and illustrated in Figure 4.4. From Figure 4.4, it can be seen that the shear force P_{sh} in a shear span is 2300 kN.

(b) Stud shear connections

When the strength of a single shear connection is D_{max}, then the number of connectors N_{ss} required in a shear span is given by

$$P_{sh} = D_{max} N_{ss} \qquad (5.3)$$

In order to determine the number of connectors in a shear span N_{ss}, we first need to determine the shear strength D_{max}.

In our beam $f_c = 25$ N/mm² (Figure 4.3), then from Eq. 5.2, $E_c = 25,250$ N/mm². From the detailing rules described in Section 5.2.2 and in order to prevent the steel flange of thickness $t_f = 18$ mm from tearing, $d_{sh} < t_f/0.4 = 45$ mm. We will, therefore, use a 19 mm diameter stud so that $A_{sh} = 284$ mm². Substituting these values into Eq. 5.1 gives $(D_{max})_{push} = 113$ kN. Therefore from Eq. 5.3, the number of connectors required in a half span is $N_{ss} = 2300/113 = 21$ connectors. It is important to realize that these dowel strengths were derived empirically from stud shear connections in which $f_u = 486$ N/mm² and, hence, it is essential to choose studs with steel strengths at least equal to those tested.

(c) Detailing

The longitudinal spacing for a uniformly distributed single line of connectors is given by $L_{si} = L_{ss}/N_{ss} = 5000/21 = 238$ mm. This is less than the maximum permitted spacing of $6h_c = 6 \times 130 = 780$ mm and greater than the minimum spacing of $5d_{sh} = 5 \times 19 = 95$ mm defined in Section 5.2.3. Therefore, these connectors could be placed in a single line but in order reduce the possibility of splitting, they will be spread over the width of the flange as in Figure 5.3(c). As $w_f = 160$ mm and $L_e = 1.3 \times 19 = 25$ mm, then $L_T = 160 - 50 - 19 = 91$ mm. As the connectors are staggered, there is no need to ensure that $L_T \geq 4d_{sh}$.

5.3.3 *Characteristic strength in composite beams approach*

5.3.3.1 Characteristic dowel strength of stud shear connectors in composite beams

The 5% characteristic strength of a group of stud shear connectors in a composite beam is given by[3]

$$\left(D_{\max}\right)_{beam} = K_{ch} A_{sh} f_u \left(\frac{f_c}{f_u}\right)^{0.35} \left(\frac{E_c}{E_s}\right)^{0.40}$$ (5.4a)

in which

$$K_{ch} = 4.7 - \frac{1.2}{\sqrt{N_{gr}}}$$ (5.4b)

and which applies within the material bounds of

$$10 < E_c < 33 \text{kN/mm}^2, \ 24 < f_c < 81 \text{N/mm}^2 \text{ and } 430 < f_u < 640 \text{N/mm}^2$$ (5.4c)

where N_{gr} is the number of connectors that can be assumed to fail as a group and which is often taken as the number of connectors in a shear span N_{ss}, and E_s is the Young's modulus for the steel used in the stud which can be taken as 200 kN/mm². The equation was derived from tests with the range of material properties in Eq. 5.4c and should not be used beyond this range.

The parameter $A_{sh} f_u$ in Eq. 5.4a represents the strength of the shank of the stud. Whereas, the parameters $(f_c/f_u)^{0.35}$ and $(E_c/E_s)^{0.40}$ are factors that cope with changes to the material properties.

Let the load/slip curves O-B-C and O-D-E in Figure 5.2 represent the behaviour of two stud shear connectors in the shear span of a composite beam. The difference between the two curves represents the normal scatter of strengths and deformations that is to be expected. Let us assume that the flexural deformation of the composite beam induces a slip in both connectors of s_z. It can be seen that the strengths of both connectors is achieved at this slip because of their plastic plateaux. Hence in composite beams with ductile connectors such as stud shear connectors, it is not necessary to design for the probability of an individual connector failing but for the group of connectors failing. This is allowed for in Eq. 5.4a by the parameter K_{ch} which is defined in Eq. 5.4b and which depends on the number of shear connectors N_{gr} that fail as a group that can be taken as the number of connectors in a shear span N_{ss}. When $N_{gr} = 1$, then $K_{ch} = 3.5$ and this represents the characteristic strength of an individual connector. Alternatively when $n \to \infty$, then $K_{ch} = 4.7$ and this represents the mean strength of the stud shear connectors.

5.3.3.2 **Example 5.2** *Characteristic strength in composite beams approach*

(a) Iterative analysis procedure
Let us redesign the beam in the previous Example 5.1 where the details of the beam are given in Section 5.3.2.2(a). An iterative procedure has to be used as the parameter K_{ch} in Eq. 5.4 depends on the number of connectors in a shear span N_{gr}, which is

initially unknown as it depends on the dowel strength which is also initially unknown. Let us use 19×100 mm studs ($A_{sh} = 284$ mm²) in which $f_u = 486$ N/mm² and $E_s = 200$ kN/mm², and which are encased in concrete with the properties $f_c = 25$ N/mm² and $E_c = 25{,}250$ N/mm². Substituting these values into Eq. 5.4 gives $D_{max} = 21{,}350K_{ch}$. The iterative procedure can be started by assuming that $(N_{gr})_1 \to \infty$, in which case from Eq. 5.4b $K_{ch} = 4.7$ and $(D_{max})_\infty = 100.3$ kN (which is the mean strength of the stud shear connectors). From Eq. 5.3, a more accurate estimate of the number of connectors is $(N_{gr})_2 = 2300/100.3 = 22.9$. Repeating the analysis but this time starting with $(N_{gr})_2 = 22.9$ gives $K_{ch} = 4.45$ and $(D_{max})_{22.9} = 95.0$ kN and $(N_{gr})_3 = 2300/95.0 = 24.2$. Now starting with $(N_{gr})_3 = 24.2$ gives $K_{ch} = 4.46$ and $(D_{max})_{24.2} = 95.2$ kN and $(N_{gr})_4 = 2300/95.2 = 24.2$, that is 25 connectors. It can be seen that the iterative procedure converges very rapidly.

If the analysis had been based on the characteristic strength of an individual connector, then in Eq. 5.4b $N_{gr} = 1$, $K_{ch} = 3.5$ and $(D_{max})_1 = 74.7$ kN and $N_{gr} = N_{ss} = 2300/74.7 = 30.8$. We would have required $30.8 - 24.2 = 6.6$ more connectors per shear span. This iterative procedure makes full use of the ductility of the shear connection and allows us to use almost the full strength of the shear connection, particularly in large beams with many connectors.

(b) Variation in the stud shear connection material properties

The dowel strength is very sensitive to the materials that comprise the stud shear connection. Let us use the lower bounds to the range of material properties given Eq. 5.4c. Therefore, the mean strength of a 19×100 mm stud of weak material strength $f_u = 430$ N/mm², that is encased in lightweight concrete of $E_c = 10$ kN/mm² of low strength $f_c = 24$ N/mm², can be derived from Eq. 5.4 as 63 kN. In contrast, using the upper bounds to the ranges of the material properties which are $f_u = 640$ N/mm², $E_c = 33$ kN/mm² and $f_c = 81$ N/mm², the mean dowel strength is 202 kN. It can be seen that the range of possible strengths for the same size of stud connection is very large.

It is also worth noting that shear strengths D_{max} greater than the tensile strength of the shank of the stud $A_{sh}f_u$ can be achieved. For example using the upper bound to the strengths in the previous paragraph, equating the shear strength of 202 kN to $A_{sh}f_u$ gives $f_u = 711$ N/mm², which is greater than the actual strength of 640 N/mm². This is because the cross-sectional area of the weld-collar shown in Figure 5.1 is about 70% greater than the cross-sectional area of the shank A_{sh} and hence the weld-collar can resist a larger shear load. Furthermore the weld collar directly resists a portion of the shear load shown as F_2 so reducing the shear force F_1 on the shank. This helps to emphasize the importance of the weld collar.

(c) Reduction in strength due to transverse ribs

The concrete slab of a composite beam is often made with steel decking or profiled sheets. The decking ribs are either in the longitudinal direction, as in Figures 2.4 and 4.8, forming a haunch around the shear connection with voids on either side of the haunch, or the ribs are in the transverse direction, as in Figures 2.6 and 4.10, encasing the shear connection in a trough with voids on either side. When the voids in the composite slab are in the vicinity of the shear connectors, they can reduce the maximum dowel strength by reducing the triaxial restraint to the concrete in the bearing zone in Figure 5.1.

The strength of the shear connection in composite beams with longitudinal ribs, that is haunches, is covered by the splitting and post-splitting analyses in Chapters 10 and 11. In contrast, the effect of transverse ribs on the maximum dowel strength of stud shear connectors is usually dealt with by applying a reduction factor R_{rib} to the maximum dowel strength D_{max}, that depends on the geometrize of both the shear connection and the composite slab. The following equation[4] is an example of a reduction factor for stud shear connectors in the troughs of composite slabs with trapezoidal profiled sheeting as in Figure 5.4.

$$R_{rib} = \frac{0.7 b_{tr}}{\sqrt{N_{tr}}\, h_{rib}} \left(\frac{h_{st}}{h_{rib}} - 1 \right) \le 1.0 \qquad (5.5a)$$

in which

$$d_{sh} \le 20 \text{mm}, f_u \le 450 \text{N/mm}^2, \text{ and } R_{rib} \le 0.8 \text{ when } N_{tr} \le 2 \qquad (5.5b)$$

where b_{tr} is approximately the mean width of the trough shown in Figure 5.4, N_{tr} is the number of connectors in each trough, and h_{rib} and h_{st} are the heights of the rib and the stud respectively.

Let us assume that the concrete element in Example 5.1 (Section 5.3.2.2(a)) is a composite slab with the transverse ribs in Figure 5.4 in which the troughs and the voids are anti-symmetric and in which $b_{tr} = 300$ mm and $h_{rib} = 70$ mm. Therefore, the longitudinal spacing of the troughs is 600 mm so that there are 5000/600 = 8 troughs in each shear span of 5 m length. From Section 5.3.3.2, the mean strength of the shear connection $(D_{max})_\infty = 100.3$ kN when $f_u = 486$ N/mm². However, Eq. 5.5 is only applicable when $f_u = 450$ N/mm² at which steel strength Eq. 5.4 gives $(D_{max})_\infty = 20,309 K_{ch} = 95.4$ kN. We would, therefore, require more than 2300/95.4 = 24 connectors as $(D_{max})_\infty$ is the mean strength not the characteristic strength. Three connectors per trough would only give 24 connectors and, hence, we would require

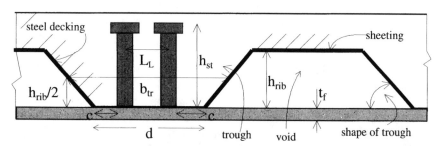

Figure 5.4 Composite beam with transverse ribs

some troughs with at least 4 connectors. From Eq. 5.5, the reduction factor with 4 connectors per trough is $(R_{rib})_4 = (0.7 \times 300)/((100/70) - 1)/(\sqrt{4} \times 70) = 0.64$.

If we used 4 connectors per trough that is a total of 32 connectors per shear span, then the strength in a shear span $P_{sh} < 0.64 \times 95.4 \times 32 = 1954$ kN (as we are using the mean strength of an individual connector), which is less than the required value for full shear connection of 2300 kN. If the slope of the trough in Figure 5.4 is 45°, then the width at the base of the trough d = 230 mm. The minimum longitudinal spacing of stud shear connectors $L_L = 5d_{sh} = 95$ mm. At this spacing, the longitudinal cover at the base of the stud c = 58 mm which is as small as one would wish. Furthermore, the width of the steel flange of 160 mm will only allow two longitudinal lines of connectors. Therefore, there is only room in each trough for 4 connectors. From Eq. 5.4b, the characteristic strength of a group of 32 connectors is less than the mean strength by a factor of 4.49/4.7 = 0.955, hence, the strength of the shear connection in a shear span $P_{sh} = 0.955 \times 1954 = 1866$ kN which is less than the required value of 2300 kN. To overcome this problem, the engineer could either change the profile, use stronger concrete, increase the height of the stud, or simply use partial-shear-connection analyses to determine the reduction in the rigid plastic flexural capacity, which would probably be slight, to see if it can be accommodated.

5.3.4 *Composite beams with non-uniform loads*
5.3.4.1 **Example 5.3** *Variable distributions of shear connectors*

Let us determine the shear connection required for the beam in Example 4.11 in Section 4.4.3.2 which is subjected to the variable loads in Figure 4.19(a). From the rigid plastic analysis in Section 4.4.3.2, we have already determined the shear flow strengths required in each part of the beam which are given in Figure 4.19(d); these were derived from the areas of each of the regions in (b). From the analysis in Figure 4.4, the required strength of the shear connection in a shear span is $P_{sh} = 2300$ kN. The strength of a 19 × 100 mm stud shear connector was derived in Section 5.3.3.2(a) as $(D_{max})_{19 \times 100} = 21{,}350K_{ch}$ where K_{ch} is defined in Eq. 5.4(b). If instead of a 19 × 100 mm stud $(A_{sh} = 284$ mm$^2)$, we will use a 16 × 80 mm stud $(A_{sh} = 201$ mm$^2)$, then from Eq. 5.4, $(D_{max})_{16 \times 80} = 21{,}350(201/284)K_{ch} = 15{,}110 K_{ch}$.

For the 16 × 80 mm stud and when n → ∞, $(D_{max})_\infty = 71.0$ kN, which from Eq. 5.3 gives a value of n = 2300/71.0 = 32.4. When $N_{ss} = N_{gr} = 32.4$, $(D_{max})_{32.4} = 71.0 \times 4.49/4.7 = 67.8$ kN and a new value of $N_{gr} = 2300/67.8 = 33.9$, that is 34 connectors per shear span; the next iteration also gives 34 connectors. Therefore, we require 34 connectors in each of the shear spans A-C and C-D in Figure 4.19. Furthermore and in shear span A-C, the 34 connectors should be distributed in proportion to the areas in (b). Therefore along the span A-B, we need 34 × (542/702) = 26.3 connectors. As shear connectors are more effective in the region near the beam supports than near the position of maximum moment, we will place 27 connectors in the span A-B and the remaining 7 in the span B-C.

In span A-B in Figure 4.19, the longitudinal spacing of a single line of connectors $L_{si} = 2500/27 = 93$ mm which is slightly more than the minimum limitation of

$5d_{sh}$ = 80 mm and considerably less than the maximum limitation of $6h_c$ = 780 mm. We could, therefore, use a single line of connectors but would prefer to space them as wide apart as possible at a maximum lateral spacing from Figure 5.3 of L_T = 160–16–42 = 102 mm and stagger them as in Figure 5.3(c). In span B-C in Figure 4.19, L_{si} = 2500/7 = 357 mm which falls well within all the limitations, and as the connectors are widely spaced we could leave them in a single line. Finally in span C-D, L_{si} =5000/34=147 mm and it is suggested that a staggered distribution be used.

5.3.5 *Composite beams designed using linear elastic theory*
5.3.5.1 General
The ultimate flexural strength of composite beams in buildings and the shear flow strengths of their shear connection are usually determined from rigid plastic analyses as described in Chapter 4. However, the serviceability behaviour of these composite beams, as described in Chapter 3, is determined from linear-elastic full-interaction analyses which are known to give satisfactory results, even though the distribution of the shear connection is based on rigid plastic analyses. In contrast, the ultimate strength of composite beams in bridges and the distribution of their connectors are both usually designed using linear-elastic full-interaction theory. The linear-elastic distribution of the shear connectors is described in this section.

5.3.5.2   **Example 5.4** *Linear elastic design of the shear connection*
Let us assume that the beam in Figure 4.3 of span of 10 m is propped during construction. Furthermore, the beam is subjected to a uniformly distributed short term load of w_{short} = 16 kN/m so that the maximum vertical shear force at the support V_{short} is 80 kN, and it is also subjected to a long term load of w_{long} = 20 kN/m that is also resisted compositely, in which case V_{long} = 100 kN. The linear-elastic properties of the components of the composite beam are given in Figure 5.5(a) where the units are in N and mm. Using the procedures described in Chapter 3, the composite beam can be transformed to a steel section with the long term and short term properties of the composite section in (b), where A_c is the area of the transformed concrete component and I_{nc} is the second moment of area of the transformed composite section. From $q = VA\bar{y}/I$ in Eq. 3.3, the maximum shear flow force at the supports due to the long term loads is $100\times10^3\times28{,}720\times(127{-}65)/716\times10^6$ = 248 N/mm, and it is 198 N/mm for the short term load, giving a total of $(Q_{sh})_{max}$ = 446 N/mm as shown at the support in Figure 5.6. Therefore, the shear flow strength required varies from 446 N/mm at the supports to zero at mid-span as shown by the line A-B.

It can be seen in Figure 4.1(b) that when there is a single line of connectors of spacing L_{si}, each shear connector resists a shear flow force within a tributary length of L_{si}. Hence, the longitudinal spacing of a single line of connectors L_{si} is given by

$$D_{max} = Q_{sh} L_{si} \qquad (5.6)$$

It would be impractical to gradually vary the spacing of the shear connectors L_{si} so that the shear flow strength followed the line A-B in Figure 5.6. Instead, the connectors are placed in blocks with a uniform spacing in each block as in Example 3.12. We could, for example, allow both a 10% overstress and a 10% understress as shown in Figure 5.6, in which case the shear flow strengths Q_{sh} required for each block of length 1 m are shown on the vertical axis. If we use for our design the mean strength of the shear connector of $D_{max} = 113$ kN, from Example 5.1 in Section 5.3.2.2(b), then the spacing of the connectors L_{si} in Figure 5.6 can be derived from Eq. 5.6. The maximum spacing permitted is $6h_c = 780$ mm, hence, this maximum spacing should be used in the region within 2 m of mid-span in Figure 5.6. Hence, the number of connectors N in each block of 1 m can be derived as shown in Figure 5.6.

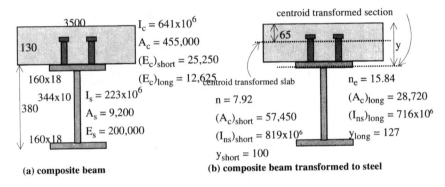

Figure 5.5 Example 5.4 Linear-elastic properties of composite beam

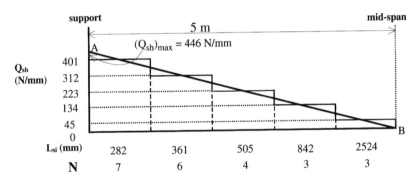

Figure 5.6 Example 5.4 Linear-elastic distribution of shear connectors

5.4 Fracture of shear connectors due to excessive slip in simply supported beams

5.4.1 *General*

It was shown in Chapter 4 how to derive the variation of the moment capacity M of a composite beam with the maximum degree of shear connection η_{max}, that is the degree of shear connection at the position of the maximum moment. A typical variation is shown in Figure 5.7 as the curve A-B-C-D. Rigid plastic analysis techniques were used to derive this failure envelope which assumed that the connectors had an unlimited slip capacity. Unfortunately, as discussed in Section 5.1 and shown in Figure 5.2, shear connectors do have a limited slip capacity, so that it is essential to ensure that fracture due to excessive slip does not occur before the design moment capacity is reached. The prevention of connector fracture is an extremely complex problem as we have to deal with the properties of the whole beam where parts may be behaving elastically whilst others are plastic. Research has shown[5] that fracture of the shear connection at a slip of S_{ult} can be represented by a curve such as G-H in Figure 5.7.

Consider a beam in which the strength of the shear connection is in excess of that required for full-shear-connection, such as at $\eta_{max} = \eta_1$ in Figure 5.7. As the applied load is increased, the moment in the beam increases along the vertical load path emanating from η_1, causing the slip in the beam to increase until the capacity M_{fsc} is reached, that depends on the weaker of the component strengths P_c or P_s as described in Chapter 4. The slip at which this moment capacity is reached is shown as s_1 in Figure 5.2 and will be less than S_p because we have more connectors than that required for full-shear-connection. Now consider a composite beam with partial-shear-connection where $\eta_{max} = \eta_2$ in Figure 5.7. The moment can be increased along the vertical load path until the capacity $M_{psc} = M_2$ is reached that depends on the strength

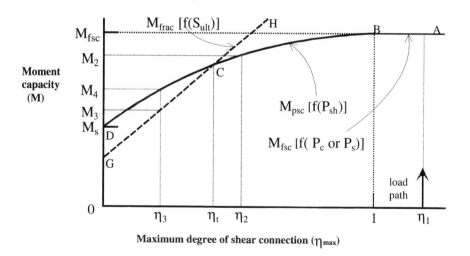

Figure 5.7 Fracture of the shear connectors in composite beams

of the shear connection P_{sh} and which occurs at the slip $s_2 < S_{ult}$ in Figure 5.2. However, if the degree of shear connection is reduced to η_3 in Figure 5.7, then fracture will occur at the moment $M_{frac} = M_3$ when $s = S_{ult}$ in Figure 2 and in which M_3 is less than the rigid plastic strength M_4.

Two methods will be described for preventing premature failure due to fracture of the shear connectors due to excessive slip in simply supported beams. The following *Parametric Study Approach*[6] is based on defining the degree of interaction η_t at the transition point C in Figure 5.7; this point defines the transition between the fracture failure envelope G-H and the rigid plastic failure envelope B-C-D. This approach is found to be in good agreement with tests in the upper regions of the degree of shear connection, that is where $\eta_{max} \geq 0.5$. In contrast, the *Mixed Analysis Approach*[5] is based on defining the fracture failure envelope G-H in Figure 5.6 and can be applied to beams with small degrees of shear connection.

5.4.2 *Slip capacities of stud shear connectors S_{ult}*

The 5% characteristic slip capacity of an individual stud shear connector that is encased in a solid concrete slab[7] is given by

$$S_{ult} = \left(0.42 - 0.0042 f_c\right) d_{sh} \tag{5.7}$$

where the units are in N and mm and where the mean slip can be derived by replacing the coefficient 0.42 with 0.48. It is felt that these slip capacities can be used for stud shear connectors in haunches (Figure 4.8) and troughs (Figure 4.10) in composite slabs as the presence of both the voids in the composite slab and the transverse reinforcement appear to make the shear connection more ductile.

Failure of a composite beam by fracture of the shear connectors due to excessive slip is very rapid and resembles an unzipping action, as the fractured connector sheds its load to adjacent connectors causing them to further slip and fracture in turn, in much the same way as the 'unbuttoning' of bolted steel connections. Therefore, the characteristic slip of an individual connector should be used in design.

5.4.3   *Parametric study approach*
5.4.3.1   Maximum slip in standard composite beams

A detailed parametric study of experimental tests and computer generated tests[6] which had standard composite sections was used to derive the following design equation.

$$S_{ult} \geq \left(\frac{M_s L h_s}{6 E_s I_s}\right)\left(\frac{L}{D}\right)^{\alpha}\left(\frac{M_{psc} - M_s}{M_s}\right)^{\beta} \tag{5.8a}$$

where

For $\eta_{max} = 0.5$, $\alpha = -0.13$ and $\beta = 1.03$ \hfill (5.8b)

For $\eta_{max} = 0.75$, $\alpha = -0.24$ and $\beta = 1.70$ \hfill (5.8c)

where M_s is the rigid plastic moment capacity of the steel component, L is the span of the beam, h_s is the height of the steel component, I_s is the second moment of area of the steel component, D is the depth of the composite beam and M_{psc} is the partial-shear-connection capacity of the composite beam.

The first parameter on the right hand side of Eq. 5.8a represents the maximum slip in the composite beam when there were no shear connectors. However, the second parameter represents the increase in slip with span and the third parameter represents the reduction in slip as the degree of shear connection increases.

5.4.3.2 **Example 5.5** *Standard composite beam*

Let us apply Eqs. 5.8 to Example 4.6 in Section 4.2.4.2 in which $\eta_{max} = 0.8$. We will assume $\eta_{max} = 0.75$ as in Eq. 5.8(c) as this will give a conservative answer. From the previous analyses, $M_s = 335$ kNm (Example 4.1 in Section 4.2.2.2(b)) and $(M_{psc})_{80\%} = 653$ kNm (Example 4.6 in Section 4.2.4.2). From Figure 4.3(b), L = 10 m, $h_s = 380$ mm and D = 510 mm and from Figure 5.5(a), $I_s = 223 \times 10^6$. Substituting these values into Eqs. 5.8(a) and 5.8(c) gives $S_{ult} > 2.13$ mm which is the maximum slip in the beam. As $f_c = 25$ N/mm^2 and $d_{sh} = 19$ mm, then from Eq. 5.7 $S_{ult} = 5.99$ mm and hence the shear connection has adequate slip capacity.

Applying Eq. 5.8(a) and 5.8(b) to the composite beam in Example 4.8 in Section 4.2.5.2 where $\eta_{max} = 0.50$ and in which $(M_{psc})_{50\%} = 569$ kNm gives $S_{ult} > 2.25$ mm, which is only slightly larger than that required with the higher degree of shear connection of 80% in the previous paragraph.

It is worth noting that the slip capacity of a stud shear connection reduces as the strength of the concrete increases, that is the connection becomes more brittle. For example, if we assume that S_{ult} in Eq. 5.7 is the maximum slip in the beam which was calculated previously as 2.13 mm, then for a 19 mm diameter stud, fracture would occur when $f_c > (0.42 - (2.13/19))/0.0042 = 73$ N/mm^2 and this reduces to 61 N/mm^2 when 13 mm studs are used.

5.4.3.3 **Example 5.6** *Strengthened composite beam*
(a) Maximum slip in the composite beam

Although Eqs. 5.8 were developed for standard composite beams with I-sections, let us apply it to the beam that was strengthened in Example 4.7 in Section 4.2.4.3 by adding a plate to the bottom flange which caused the degree of shear connection to reduce to 50%. The strengthened beam is shown in Figure 4.14 and the ultimate strength is determined in Section 4.2.4.3. For this strengthened beam: the rigid plastic moment capacity of the plated steel component can be determined from a similar analysis to that depicted in Figure 4.5 except that the neutral axis in (c) now occurs at the bottom flange plate interface in Figure 4.14(a), from which taking moments about the top fibre of the steel beam gives $M_s = (2288 \times 0.393) - (720 \times 0.371) - (860 \times 0.190) - (720 \times 0.009) = 462$ kNm; L = 10 m; $h_s = 406$ mm; $(M_{psc})_{50\%} = 1500$ kNm; D = 536 mm; and the second moment of area of the plated steel component $I_s \approx (160 \times 18 \times 259^2) + (10 \times 344^3/12) + (10 \times 344 \times 78^2) + (160 \times 18 \times 103^2) + (220 \times 26 \times 125^2) = 368 \times 10^6$ mm^4, where the neutral axis is 268 mm from the top of the steel beam. Substituting into Eqs. 5.8(a) and

5.8(b) gives $S_{ult} > 6.68$ mm which is greater than the characteristic slip of the shear connection being used of 5.99 mm (which was derived in Section 5.4.3.2). Therefore, the shear connection will fracture and prevent the flexural capacity of 1500 kNm being achieved.

(b) Flexural capacity at fracture of connectors

As fracture will occur prematurely, we can use Eq. 5.8 to estimate the maximum moment capacity that can be achieved, by substituting the actual slip capacity of 5.99 mm for S_{ult} into Eq. 5.8 gives $5.99 = 4.248 \times 0.684 \times ((M_{ps} - 462.2)/462.2)^{1.03}$. Solving gives $(M_{psc})_{50\%} = 1{,}396$ kNm which is still substantially stronger than the unplated composite capacity of 702 kNm.

5.4.4 *Mixed analysis approach*

5.4.4.1 Uniform distribution of shear connectors and a uniformly distributed applied load

The mixed analysis approach is an exact solution for an idealized composite beam in which the steel and concrete components remain linear-elastic but the shear connector component is fully plastic[5]. The maximum slip S_{max}, in a simply supported composite beam of span L, with a uniform distribution of shear connectors of strength P_{sh} per shear span, and subjected to a uniformly distributed applied load that induces a maximum moment M_{max} is given by

$$S_{max} = \frac{M_{max}L}{3}K_1 - \frac{P_{sh}L}{4}K_2 \tag{5.9}$$

in which

$$K_1 = \frac{h_{cent}}{E_cI_c + E_sI_s} \tag{5.10}$$

and

$$K_2 = \frac{h_{cent}^2}{E_cI_c + E_sI_s} + \frac{1}{E_cA_c} + \frac{1}{E_sA_s} \tag{5.11}$$

where h_{cent} is the distance between the centroid of the concrete component and the centroid of the steel component.

The first parameter on the right hand side of Eq. 5.9, that is $(M_{max}LK_1/3)$ is the maximum slip in the composite beam without shear connectors, whereas, the second parameter $(P_{sh}LK_2/4)$ is the reduction in slip due to the shear connectors. It is worth noting that the first parameter depends on the applied moment, while the second parameter depends on the shear connector forces.

5.4.4.2 **Example 5.7** *Standard composite beam with a uniformly distributed load*

Let us determine the maximum slip that occurs at the rigid plastic moment capacity of the composite beam in Figure 4.3 when: the beam has a 50% degree of shear connection; the shear connectors are uniformly spaced along the beam; and the beam is subjected to a uniform distribution of the applied load. The elastic properties of the steel and concrete components of the composite beam are given in Figure 5.5(a).

The variables in Eq. 5.10 are (from Figure 4.3) $h_{cent} = (130/2) + (380/2) = 255$ mm and (from Figure 5.5) $I_s = 223 \times 10^6$ mm^4, $I_c = 641 \times 10^6$ mm^4, $E_s = 200$ kN/mm^2 and $E_c = 25,250$ N/mm^2. Hence, $K_1 = 26 \times 125^2$ (Nmm)$^{-1}$. The remaining variables in Eq. 5.11 are (from Figure 5.4) $A_c = 455,000$ mm^2 and $A_s = 9,200$ mm^2 which gives $K_2 = 1.70 \times 10^{-9}$ N^{-1}. From Example 4.8 in Section 4.2.5.2, $M_{max} = (M_{psc})_{50\%} = 569$ kNm and $P_{sh} = 2300/2 = 1150$ kN and from Figure 4.3 L = 10 m. Substituting these values into Eq. 5.9 gives $S_{max} = 7.97 - 4.89 = 3.08$ mm. Therefore, the maximum slip in the stud shear connectors which occurs at the supports is 3.08 mm. It is worth noting that the maximum slip with 'no shear connectors', which is the first parameter in Eq. 5.9, is 7.97 mm which would easily fracture a 19 mm stud shear connector at each support. However, the compressive force induced in the concrete component by the shear connectors has reduced this slip by 4.89 mm, which is the second parameter in Eq. 5.9, to give an overall slip of 3.08 mm which is unlikely to fracture 19 mm stud connectors.

5.4.4.3 *Uniform distribution of shear connectors and a point load applied at mid-span*

The maximum slip in a composite beam also depends on the type of applied load. For the case of a point load applied at mid-span (instead of the uniformly distributed load in the previous example) on a beam which still has a uniform distribution of shear connectors, the maximum slip is given by

$$S_{max} = \frac{M_{max} L}{4} K_1 - \frac{P_{sh} L}{4} K_2 \tag{5.12}$$

Comparing Eq. 5.9, for a uniformly distributed load, with Eq. 5.12, for a concentrated load at mid-span, it can be seen that the uniformly distributed load is the critical case, as a larger slip is induced when there are no shear connectors, that is the first parameter in Eq. 5.9 is larger than the first in Eq. 5.12. It is worth noting that the second parameter in both equations is the same, as this depends on the strength and distribution of the shear connectors which is obviously the same in both beams.

5.4.4.4 **Example 5.8** *Standard composite beam with a point load*

Applying Eq. 5.12 to the composite beam with the properties in Section 5.4.4.2 above gives $S_{max} = 7.97 \times (3/4) - 4.89 = 1.09$ mm, where the coefficient (3/4) is the ratio of the first parameters in Eqs. 5.9 and 5.12. The slip of 1.09 is considerably less than the slip of 3.08 mm for a uniformly distributed load. Hence, if the connectors are unlikely to fracture in a composite beam that subjected to a uniformly distributed load, then there is no need to check for fracture when point loads are applied.

5.4.4.5 **Example 5.9** *Strengthened composite beam*
(a) Maximum slip in composite beam

Let us apply this mixed analysis approach to the strengthened composite beam in Example 4.7 which is described in Section 4.2.4.3 and shown in Figure 4.7 and which has already been analysed in Section 5.4.3.3 using the parametric-study approach. From Section 5.4.3.3, $I_s = 368 \times 10^6$ mm^4 (where the centroid is 268 mm from the top fibre of the steel component), therefore, $h_{cent} = (130/2) + 268 = 333$ mm, and from Figure 5.5(a), $E_s = 200$ kN/mm^2, $A_c = 455 \times 10^6$ mm^2, $I_c = 641 \times 10^6$ and $E_c = 25,250$ N/mm^2 and from Figure 4.7 $A_s = 14,920$ mm^2. Hence from Eq. 5.10, $K_1 = 3.70 \times 10^{-12}$ (Nmm)$^{-1}$ in Eq. 5.10 and $K_2 = 3.70 \times 10^{-12} \times 333 + 8.70 \times 10^{-11} + 3.35 \times 10^{-11} = 1.655 \times 10^{-9}$ N^{-1} in Eq. 5.11. Furthermore, from Section 4.2.4.3 $M_{max} = (M_{psc})_{50\%} = 1,500$ kNm, $P_{sh} = 2300$ kN and L = 10 m. Applying Eq. 5.9 (for the case of a uniformly distributed load) gives $S_{max} = 18.5 - 9.5 = 9.0$ mm. Therefore, fracture will occur before the rigid plastic moment capacity of 1,500 kNm is reached as the characteristic slip capacity, which was derived in Section 5.4.3.2, is only 5.99 mm.

(b) Flexural capacity at fracture of connectors

From the analysis in (a) above, it was shown that fracture will occur before the rigid plastic capacity is reached. Therefore, the fracture failure envelope, such as G-H in Figure 5.7, is controlling the strength of the composite beam. This failure envelope, which is given by Eq. 5.9 for a uniformly distributed load, will now be used to predict the capacity of the composite beam. Substituting into Eq. 5.9 $S_{max} = 5.99$ mm, which is the slip capacity of the stud shear connectors, and using the sectional and material properties in (a) above, gives $M_{max} = ((S_{max} + P_{max}LK_2/4)3)/LK_1 = (5.99 + 9.52) \times 81 \times 10^6 = 1,256$ kNm which is the moment capacity of the composite beam.

5.4.4.6   *Variable distributions of both the shear connectors and the applied loads*

Equations 5.9 and 5.12 are specific to a type of loading and to a beam with a uniform distribution of the shear connectors. However, these equations can be written in the following generic form that can be applied to any distribution of shear connectors and to any distribution of the applied load.

$$\left(S_{max}\right)_\alpha = \left(A_m\right)_\alpha K_1 - \left(A_{sr}\right)_\alpha K_2 \qquad (5.13)$$

where $(S_{max})_\alpha$ is the maximum slip in the shear span designated α, and where $(A_m)_\alpha$ is the area of the moment diagram in the shear span α and $(A_{sr})_\alpha$ is the area of the longitudinal shear force or thrust diagram in the shear span α.

The area-parameters $(A_m)_\alpha$ and $(A_{sr})_\alpha$ in Eq. 5.13 are shown diagrammatically in Figure 5.8 for a composite beam with eight groups of connectors which are shown at discrete points in each shear span. The connectors impose a thrust on the concrete component which is zero at the supports and accumulates along the length of the shear span. This is shown as starting at F_1 in the right hand shear span in (a) and finishing at the total thrust of P_{sh} at the maximum moment position. The distribution of the longitudinal thrust in each shear span is plotted in (b), where the area of the diagram in a shear span is equal to the longitudinal thrust parameter $(A_{sr})_\alpha$. The composite beam in (a) is subjected to a variable load which causes the distribution of the applied moment in (c), where the applied moment parameter $(A_m)_\alpha$ is the area in a shear span.

As can be seen in Eq. 5.13, the slip at the end of a beam is a function of the difference between A_m in Figure 5.8(c) and A_{sr} in (b). This difference is usually greatest in the longer shear span, however, the difference also depends on the distribution of the shear connectors. For example, if all the connectors are concentrated at a support, such as in the right hand shear span in (a), then $(A_{sr})_r$ in (b) would be rectangular, that is at its maximum value, so that the slip from Eq. 5.13 will be at its least value. Needless to say, it would be foolish to concentrate all the connectors at the supports, as the structure would no longer be acting as a composite beam but more like an arch.

5.4.4.7 **Example 5.10** *Variable distribution of shear connectors and loads*

Let us analyse the composite beam in Example 5.7 in Section 5.4.4.2 where all of the properties are give in Section 5.4.4.2 and from which $K_1 = 4.20 \times 10^{-12}$ $(Nmm)^{-1}$

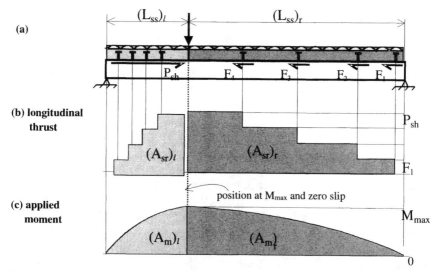

Figure 5.8 Variable loads and connector distributions

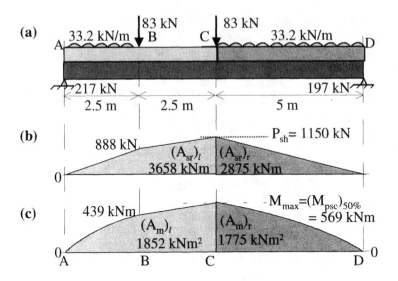

Figure 5.9 Example 5.10 Slip in shear spans

and $K^2 = 1.70 \times 10^{-9}\,N^{-1}$ was calculated. The beam has a moment capacity of $(M_{psc})_{50\%}$ = 569 kNm and, hence, can just resist the distribution of applied loads in Figure 5.9(a) in which M_{max} = 569 kNm as shown in (c). Integrating the applied moment diagram over region A-B in (c) gives

$$\left(A_m\right)_{A-B} = \int_0^{2.5} 217x - 33.2x^2 / 2 = 592 \, \text{kNm}^2$$

and integrating over B-C gives $(A_m)_{B-C} = ((439 + 568)/2) \times 2.5 = 1260\,\text{kNm}^2$, therefore, $(A_m)_{A-C} = 592 + 1260 = 1852\,\text{kNm}^2$ as shown in (c). Integrating over C-D in (c) gives $(A_m)_{C-D} = 1775\,\text{kNm}^2$.

The beam has 50% shear connection so that the strength of the shear connection in a shear span, which is also the thrust of the connectors at the position of maximum moment, is P_{sh} = 2300/2 = 1150 kN, as shown at the mid-span in Figure 5.9(b). The distribution of the applied loads in Figure 5.8(a) was chosen so that they are exactly in the same proportion to those in Figure 4.19(a). Therefore, the area of the vertica shear force distribution of the beam in Figure 5.9(a) has the same proportions as those in Figure 4.19(b). Hence, Figure 4.19(b) can be used to distribute the connectors in the shear span A-C in Figure 5.9(a). From Figure 4.19(b), the proportion of connectors in A-B is 542/ (542 + 160) = 0.772. Therefore the strength of shear connection required in A-B in Figure 5.9(a) is 0.772 × 1150 = 888 kN which is the thrust at section B in (b). As there are a fairly large number of connectors, we will ignore their discrete positions and assume the linear variation in thrust shown which gives from their areas $(A_{sr})_l = (888 \times 2.5 / 2) + (888 + 1150) \times 2.5/2 = 3658\,\text{kNm}^2$ and $(A_{sr})_r = 1150 \times 5/2 = 2875$ kNm.

Applying Eq. 5.13 gives the maximum slip in the right shear span of $(S_{max})_r =$ $(2875 \times 10^9 \times 4.2 \times 10^{-12}) + (2875 \times 10^6 \times 1.70 \times 10^{-9}) = 7.45 - 4.89 = 2.56$ mm, and that at the left of $(S_{max})_l = 7.78 - 6.22 = 1.55$ mm. It can be seen that the concentration of connectors in region A-B in Figure 5.8(b) has made the slip on the left hand side much less than that in the right hand side.

5.4.4.8 *Fracture failure envelope*
Equation 5.9, which is based on a uniformly distributed applied load and a uniform connector distribution, can be re-arranged to give the following fracture failure envelope

$$M_{max} = \frac{3}{LK_1}\left(S_{max} + \frac{P_{sh} LK_2}{4}\right) \tag{5.14}$$

As the maximum degree of shear connection η_{max} is directly proportional to the strength of the shear connection P_{sh}, it can be seen that the fracture failure envelope defined by Eq. 5.14 is linearly proportional to the degree of shear connection.

5.4.4.9 **Example 5.11** *Failure envelopes for long and short term loads*
Let us consider the beam in Figure 4.3 in which there is a uniform distribution of shear connectors and which is subjected to a uniformly distributed load. We will assume that the beam has 13 mm diameter stud shear connectors of $S_{ult} = 4.10$ mm.. The flexural capacity of this beam has already mean analysed for different degrees of shear connection and the results are plotted as the rigid plastic failure envelope in Figure 5.10.

The elastic properties of the beam are given in Figure 5.5. Furthermore, the beam has already been analysed for fracture due to short term loads in Section 5.4.4.2 where the values for $K_1 = 4.20 \times 10^{-12}$ (Nmm)$^{-1}$ and $K_2 = 1.70 \times 10^{-9}$ N^{-1} have been used in Eq. 5.14 to plot the short term fracture enveloped in Figure 5.9. It can be seen that the fracture failure envelope governs the strength at low degrees of shear connection in region A-B, whilst the rigid plastic failure envelope governs the strength at the higher degrees of shear connection in the region B-C. If the rigid-plastic linear-interpolation envelope is used, then the fracture failure envelope only governs over the smaller region A-D as the interpolation method under-estimates the strengths.

From the long term properties in Figure 5.4, $K_1 = 4.84 \times 10^{-12}$ and $K_2 = 7.18 \times 10^{-10}$, from which can be derived the long term fracture failure envelope in Figure 5.10. This envelope can be seen to control the strength for all degrees of shear connection. As beams are never designed to withstand the maximum design loads for long periods, the true fracture failure envelope will lie between the two extremes (that is between the long and short term fracture envelopes) in Figure 5.10, and should approach the short term fracture failure envelope.

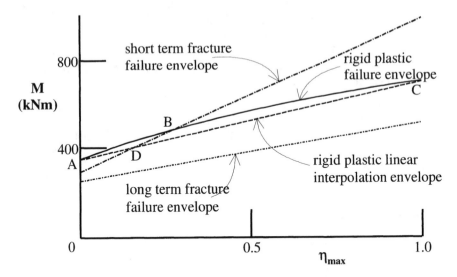

Figure 5.10 Failure envelopes

5.5 References

1. Oehlers, D.J. and Bradford, M.A. (1995). Composite Steel and Concrete Structural Members: Fundamental Behaviour. Pergamon Press, Oxford.
2. Ollgaard, J.G., Slutter, R.G. and Fisher, J.W. (1971). 'Shear strength of stud shear connectors in lightweight and normal-density concrete'. Engineering Journal, American Institute of Steel Construction, Vol.8, 55-64.
3. Oehlers, D.J. and Johnson, R.P. (1987). 'The strength of stud shear connections in composite beams.' The Structural Engineer, Part B, June, Vol. 65B, Number 2, 44-48.
4. Johnson, R.P. (1994). Composite Structures of Steel and Concrete. 2nd edn, Blackwood Scientific Publications, U.K.
5. Oehlers, D.J. and Sved, G (1995). 'Flexural strength of composite beams with limited slip capacity shear connectors.' Journal of Structural Engineering, ASCE, June, Vol. 121 No. 6, 932-938.
6. Johnson, R.P. and Molenstra, N. (1991). 'Partial shear connection in composite beams in buildings.' Proceedings, Institution of Civil Engineers, London, Part 2, Vol. 91, 679-704.
7. Oehlers, D.J. and Coughlan, C.G. (1986). 'The shear stiffness of stud shear connections in composite beams'. Journal of Constructional Steel Research, Vol. 6, 273-284.

6 Transfer of longitudinal shear forces

6.1 Introduction

It has been shown in Chapters 3 and 4 that the composite action between the steel and concrete components of a composite beam induces longitudinal shear forces in the mechanical shear connectors that tie these components together. These longitudinal shear forces are transferred into the concrete component by the dowel action of the mechanical shear connectors, as described in Chapter 5, and then these concentrated dowel forces are dispersed longitudinally and laterally into the concrete component. This chapter deals with the ability of the concrete component to resist the longitudinal component of this dispersal, whereas the lateral component of the dispersal is dealt with in Chapter 10.

It is convenient in the analysis of the longitudinal shear transfer to convert the longitudinal shear forces into shear flow forces, that is the longitudinal force per unit length of beam q. In designing the concrete component to resist the shear flow forces, it is first necessary to identify the critical longitudinal planes in which failure can occur and these planes are referred to in the following section as the shear flow planes. The next step is to quantify the shear flow forces in these shear flow planes and then to ensure that the resistance of these planes, that is their shear flow strengths, is adequate.

6.2 Shear flow planes

The concrete component of a composite beam is shown in Figure 6.1(a). The hatched region at A is a longitudinal shear plane of area $L \times L_p$ that is subjected to a total uniformly distributed longitudinal shear force of H. For a the length of the shear plane L of unit length as shown, the shear force is H/L which is now a shear flow force, that is a force per unit length, which is shown as q_A. Furthermore, the area of the shear plane is now $(L_p)_A \times 1 = (L_p)_A$ which will be referred to as the perimeter length. Hence when dealing with shear flows, the perimeter length L_p defines a shear plane that extends a longitudinal distance of one unit.

There are an infinite number of shear planes and it is necessary to use engineering judgement to determine the critical planes that are most likely to fail or to govern the design. These planes can be categorized as those that traverse the depth of the concrete component such as $(L_p)_A$, $(L_p)_B$ and $(L_p)_E$ in Figure 6.1 and those that encompass the connectors such as $(L_p)_C$, $(L_p)_D$, $(L_p)_F$, $(L_p)_G$ and $(L_p)_H$.

6.3 Shear flow forces

The shear flow force imposed by the connectors on the concrete component is shown in Figure 6.1 as q_t. The shear flow force q_t can be derived from Chapter 3 or

Chapter 4 depending on whether the analysis is elastic or plastic. Alternatively, q_t can be derived directly from the connector spacings in Chapter 5, if it is assumed that the connectors are fully loaded. For example, if the connector spacings at a design point in a beam are those shown in Figure 5.3(b), then q_t is the strength of a single connector D_{max} divided by the longitudinal spacing when the connectors are placed in a single longitudinal line L_{si}. In both Figures 5.3(b) and (c), $L_{si} = L_L/2$ even though the connectors are staggered in the latter example (it may be worth referring to Eq. 5.6 which is also based on the concept of deriving the shear flow from a single line of connectors). The shear flow force q_t in Figure 6.1(a) is gradually dispersed into the concrete component, so that the force in the shear planes L_p diminishes with distance from the connectors, reducing to zero at the sides of the concrete component. The next step in the analysis is to determine the shear flow force at each critical shear flow plane such as q_A at $(L_p)_A$.

The shear flow force q_t that is shown adjacent to the shear connectors in Figure 6.2 is in equilibrium with the compressive force along the whole section e-f-g-h, which is shown as uniformly distributed as the effects of shear lag are ignored, so that e-f-g-h can be considered to be the effective width of the slab as described in Section 2.2.2. The shear flow force along any longitudinal plane can be determined by cutting the concrete

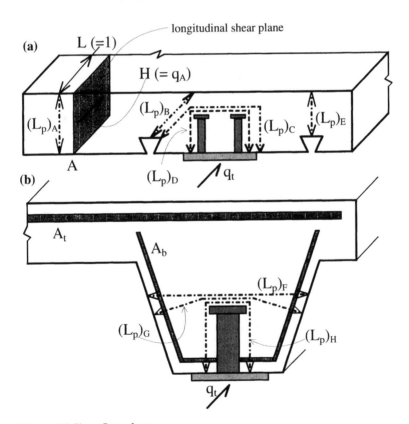

Figure 6.1 Shear flow planes

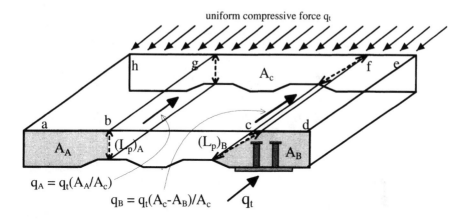

Figure 6.2 Shear flow forces

component along the plane and determining the resultant force that acts on either of the free bodies. For example, the resultant force on the shear plane represented by $(L_p)_A$ is the resultant force on the free body a-b-g-h, which is the compressive force acting over the area A_A. If A_c is the total cross-sectional area of the concrete component, then q_A is the proportion A_A/A_c of q_t as shown. For shear plane $(L_p)_B$, the resultant force on the free body c-d-e-f is q_t minus the compressive force that acts over the area A_B, that is $q_t A_B/A_c$. Alternatively, the shear force on $(L_p)_B$ is the resultant force on the free body a-c-f-h which is shown in Figure 6.2 as q_B.

6.4 Generic shear flow strengths

Fundamental research[1,2] has shown that the weakest shear planes occur where there is a longitudinal crack in the concrete which may have been formed by transverse flexure or longitudinal splitting as described in Chapter 10 and, hence, the strength of a cracked shear plane is generally used in design. The mechanism by which shear is transferred across a crack is shown in Figure 6.3. The shear is resisted by the dowel action of the transverse reinforcement which is shown as a bend in the reinforcement in Figure 6.3. The longitudinal shear forces F induce slip between the two crack surfaces which causes the two elements to separate as the aggregate particles on each surface ride over each other. This separation, due to the shear displacement, stretches the reinforcing bars and induces passive tension in the bars which is balanced by the passive compressive forces across the crack which in turn resist the shear by passive friction; this mechanism is often referred to as aggregate interlock. It can, therefore, be seen that the transverse reinforcement is essential to the transfer of shear across the shear plane by both dowel action and by aggregate interlock action. Shear is also transferred across the interface by friction imposed by direct compression across the interface which is shown as σ_{nf}.

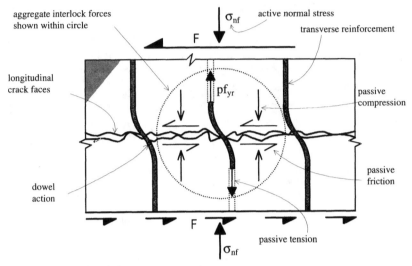

Figure 6.3 Transfer of shear across a crack

The following characteristic strength $(v_u)_{char}$ of a cracked shear plane was determined directly by tests and is dependent on: the tensile strength of the concrete f_{ct}; the axial strength of the reinforcement per unit area of the shear plane pf_{yr} which is the yield strength when the reinforcement is fully anchored or the bond strength when the reinforcement is not fully anchored; and the active normal stress σ_{nf} which is positive when compressive and negative when tensile.

$$\left(v_u\right)_{char} \; = \; 0.66 f_{ct} + 0.8 pf_{yr} + 0.8 \sigma_{nf} \tag{6.1}$$

For convenience, Eq. 6.1 can be written in terms of shear flow strengths of a shear plane

$$Q_{ch} \; = \; 0.66 f_{ct} L_p + 0.8 A_{tr} f_{yr} + 0.8 F_{nf} \tag{6.2}$$

where A_{tr} is the area of the transverse reinforcement per unit length of beam, f_{yr} is the yield strength of the transverse reinforcement when fully anchored or the maximum stress that can be achieved in the reinforcement when it is not fully anchored, and F_{nf} is the normal force per unit length. Tests have shown that the parameter $0.66 f_{ct} L_p$ requires a minimal active or passive normal force and, hence, Eq. 6.2 has a lower bound of

$$0.8\left(A_{tr} f_{yr} + F_{nf}\right) \; \geq \; 0.53 f_{ct} L_p \tag{6.3}$$

Furthermore, tests have also shown that the mechanism of shear resistance portrayed in Figure 6.3 has an upper bound beyond which the cracked section behaves as uncracked, and this upper bound limit is given by

$$Q_{ch} \leq 0.3 f_c L_p \tag{6.4}$$

6.5 Resistance of shear plane traversing depth of slab

6.5.1   *Shear flow strength of full depth plane*

Examples of full depth shear planes are shown in Figure 6.1(a) as $(L_p)_A$, $(L_p)_B$ and $(L_p)_E$ and the stress resultants acting on these planes are shown in Figure 6.4. Shear is resisted by the contribution of the top and bottom transverse reinforcement of areas A_t and A_b to the aggregate interlock and dowel action mechanisms. Furthermore as the shear plane traverses the full depth of the slab, the resultant normal force acting on the shear plane F_{nf} due to transverse flexure is zero. Therefore from Eq. 6.2, the shear resistance of a full depth plane is given by

$$Q_{ch} = 0.66 f_{ct} L_p + 0.8 \left(A_t + A_b \right) f_{yr} \tag{6.5}$$

and the lower bound of Eq. 6.3 becomes

$$0.8 A_{tr} f_{yr} \geq 0.53 f_{ct} L_p \tag{6.6}$$

and the upper bound of Eq. 6.4 remains unchanged.

Example 6.1 *T-beam with solid slab*

The concrete component of the composite T-beam in Figure 4.3 is shown in Figure 6.5. The stud shear connectors have already been designed in Example 5.1 in Section 5.3.2.3 and the results are summarized on the right hand side of the figure. Units of N and mm are used throughout unless shown otherwise.

Let us consider the shear plane A-A in Figure 6.5. The perimeter length of the shear plane $(L_p)_A = 130$ mm. The shear flow strength of the shear plane is given by Eq. 6.5 in which the direct tensile strength of the concrete f_{ct} can be derived from the following relationship with the cylinder compressive strength f_c as given in Section 1.3.5.1.

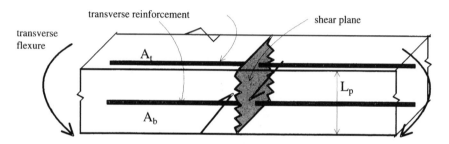

transverse reinforcement   shear plane

transverse flexure

A_t

L_p

A_b

Figure 6.4 Full depth shear planes

$$f_{ct} = 0.4\sqrt{f_c} \qquad\qquad (6.7)$$

in which the units are in N and mm and when the concrete is of normal density. Hence, $f_{ct} = 2\ \text{N/mm}^2$. The shear flow force imposed by the connectors $q_t = D_{max}/L_{si} = 113{,}000/238 = 475\ \text{N/mm}$. The area of the free body to the left of the shear plane $A_A = 221{,}325\ \text{mm}^2$, hence $q_A = (A_A/A_c)q_t = 0.49q_t$. For all intents and purposes $q_A = (1750/3500)q_t = q_t/2 = 237\ \text{N/mm}$, that is the shear flow force is in proportion to the width of the free body as a proportion of the effective width of the concrete component.

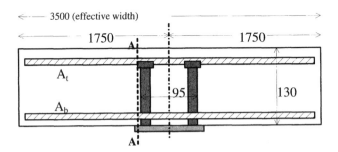

All units in N and mm unless stated:
$f_c = 25\ \text{N/mm}^2$
$D_{max} = 113\ \text{kN}$
$L_{si} = 238\ \text{mm}$
$A_c = 455{,}000\ \text{mm}^2$
$f_{yr} = 400\ \text{N/mm}^2$

Figure 6.5 T-beam with solid slab

The area of transverse reinforcing bars required to resist q_A can be derived from Eq. 6.5 by substituting q_A for Q_{ch} which gives $A_{tr} = A_b + A_t = (237 - 0.66\times2\times130)/(0.8 \times 400) = 0.203\ \text{mm}$, which is the area required per unit length of shear plane. However, the lower bound requirement of Eq. 6.6 gives $A_{tr} > (0.53\times2\times130)/(0.8\times400) = 0.431\ \text{mm}$ and, hence, this condition controls the design. Using 8 mm diameter bars of yield strength $f_{yr} = 400\ \text{N/mm}^2$ would require a longitudinal spacing of these tansverse bars of $L_{tr} = (\pi\times8^2/4)\,/\,0.431 = 117\ \text{mm}$. It is usually assumed that the transverse reinforcement in the top of the slab, such as A_t in Figure 6.5, that is required to resist the hogging or negative moment in the slab over the composite beam can also resist the longitudinal shear. Furthermore the bottom transverse reinforcement A_b is often part of the sagging reinforcement in the slab that is extended over the supports, that is over the composite beam. It is worth noting that the upper bound requirement of Eq. 6.4 will allow a shear flow force of $0.3\times25\times130 = 975\ \text{N/mm}$ which is substantially larger than the requirement of 237 N/mm

6.5.1.2 **Example 6.2** *Longitudinally spanning profiled sheets*
Let us replace the solid concrete slab in Figure 6.5 with the profiled slab in Figure 6.6 in which the ribs of the trapezoidal sheeting are parallel with the longitudinal span of the composite beam and in which the composite beam is slightly eccentric to the slab. By inspection, a critical shear flow plane will occur at B-B where the depth of the slab is reduced by the rib of the profiled sheeting. Hence, $(L_p)_B = 70\ \text{mm}$. The shear flow force exerted by the shear connectors q_t remains at 475 N/mm.

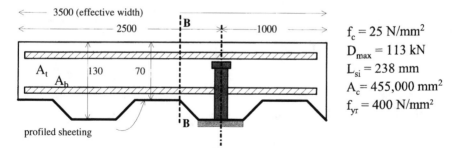

Figure 6.6 Profiled slab with longitudinal ribs

As the shear plane B-B will be assumed to be close to the shear connectors, $q_B \approx q_t(2500/3500) = 339$ N/mm which is less than the upper bound value of $0.3 \times 25 \times 70 = 525$ N/mm from Eq. 6.4. Applying Eq. 6.5 gives $A_{tr} = (339 - 0.6 \times 2 \times 70)/(0.8 \times 400) = 0.797$ mm which is larger than the minimum requirement of $(0.53 \times 2 \times 70)/(0.8 \times 400) = 0.231$ mm from Eq. 6.6.

The profiled sheeting in Figure 6.6 crosses the shear plane and, therefore, may be expected to contribute to the longitudinal shear strength of the shear plane. However in order to contribute to the longitudinal shear strength, it must be able to provide passive tension across the shear plane as shown in Figure 6.3. It can be seen in Figure 6.6 that any transverse tension in the profiled sheeting will simply straighten the sheeting and cause it to unravel from the concrete slab. Hence, the profiled sheeting is not anchored in the transverse direction and, therefore, does not contribute to the longitudinal shear strength. All of the required transverse reinforcement of $A_{tr} = 0.797$ mm must, therefore, be supplied by transverse reinforcing bars; this would require 8 mm diameter bars at $(\pi 8^2/4)/0.797 = 63$ mm centres which can be placed both at the top of the slab or just above the profiled sheeting as shown in Figure 6.6.

6.5.1.3 **Example 6.3** *Transverse spanning profiled sheets*
The composite slab with longitudinal trapezoidal ribs in Figure 6.6 has been replaced with a composite slab with dove-tailed transverse ribs in Figure 6.7. In this case, the ribs can resist transverse tension and, therefore, will contribute to the longitudinal shear strength.

(a) Shear plane supporting short shear span
Let us first consider the shear plane at C-C in Figure 6.7(a) and the strength of the sheeting to the right of this section. The cross section of the sheeting is shown in Figure 6.7(b). The ribs are at 200 mm centres and the area of sheeting for each 200 mm width can be derived as $50 + 67 + 67 + 200 = 384$ mm² from the dimensions given. Therefore, the area of profiled sheeting per unit longitudinal length $A_{prof} = 384/200 = 1.92$ mm and the yield strength of the profile sheet per unit length $A_{prof}f_{yp} = 1.92 \times 550 = 1056$ N/mm. However, this yield strength can only be achieved if the sheeting is fully anchored. Let us assume that the bond strength of the sheeting $f_b = 0.25$ N/mm². This bond strength acts over

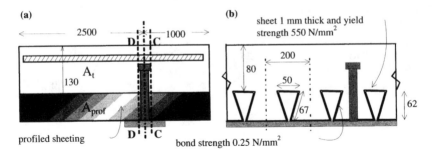

Figure 6.7 Profiled slab with transverse ribs

the perimeter of the rib of length $50 + 67 + 67 = 184$ mm and over the length of the rib of 1000 mm. Hence the bond strength per rib $= 184 \times 1000 \times 0.25 = 46,000$ N which over a unit longitudinal length is $46,000/200 = 230$ N/mm. As the bond strength of 230 N/mm is less than the yield strength of 1056 N/mm, the bond strength controls the design so that the effective strength of the profiled sheeting '$A_{prof} f_{yp}$' = '$A_{tr} f_{yr}$' = 230 N/mm.

The voids in the concrete slab encased by the transverse dove-tailed ribs in Figure 6.7(b) will reduce the ability of the concrete to transfer the shear. This reduction in strength can simply be catered for by using an effective depth of slab, which is the depth of a solid slab with the same cross-sectional area of concrete as the profiled slab. In this example, the effective depth is $((200\times130) - (50\times62/2)) \times 130 / (200\times130)$ $= 122$ mm which is only slightly smaller than the overall depth of 130 mm, however the reduction can be substantial when transverse trapezoidal ribs are used. Hence, the perimeter length $(L_p)_C = 122$ mm. The shear flow force imposed by the connectors q_t remains unchanged from the previous examples at 475 N/mm so that $q_C = q_t(1000/3500) = 136$ N/mm. In order to determine the area of transverse reinforcement required, Eq. 6.5 can be written as the following equation with the values of the parameters shown immediately below.

$$
\begin{array}{cccc}
q_C & = & 0.66 f_{ct} L_p + 0.8 A_{prof} f_{yp} + 0.8 A_{tr} f_{yr} \\
136 & & 161 \qquad\qquad 184 \qquad\qquad 0.8 A_{tr} f_{yr}
\end{array}
$$

It can be seen that the parameter $0.66 f_{ct} L_p = 161$ N/mm is sufficient to resist the shear flow force of $q_C = 136$ N/mm. Therefore, only the minimum transverse reinforcement is required which from Eq. 6.6 comes to $A_{tr} f_{yr} > (0.53\times2\times122)/0.8 = 162$ N/mm which is less than that already supplied by the profiled sheeting of '$A_{prof} f_{yp}$' = 230 N/mm. These calculations show that the shear flow force can be transferred into the cantilevered section without the addition of transverse reinforcing bars.

(b) Shear plane supporting long shear span
Let us now consider shear plane D-D in Figure 6.7(a). The shear flow force $q_D = 475 \times (2500/3500) = 339$ N/mm. The bond strength of the profiled sheeting at D-

D is virtually the same as that at C-C as it is governed by the shorter of the two bond lengths, of approximately 1 m and 2.5 m, which remains virtually unchanged at close to 1000 mm. Applying Eq. 6.5 gives

$$q_D = 0.66 f_{ct} L_p + 0.8 A_{prof} f_{yp} + 0.8 A_{tr} f_{yr}$$
$$339 \qquad 161 \qquad\qquad 184 \qquad\qquad 0.8A_{tr}f_{yr}$$

As the first two terms on the right hand side just exceed the shear flow force, the shear flow plane is sufficiently strong.

If the cantilevered section in Figure 6.7 is halved from 1 m to 0.5 m, then $q_D = 475 \times (3000/3500) = 407$ N/mm and the bond strength '$A_{prof} f_{yp}$' is halved to 115 N/mm. Equation 6.5 now becomes

$$q_D = 0.66 f_{ct} L_p + 0.8 A_{prof} f_{yp} + 0.8 A_{tr} f_{yr}$$
$$407 \qquad 161 \qquad\qquad 92 \qquad\qquad 0.8A_{tr}f_{yr}$$

from which $A_{tr} = 0.481$ mm which would require 8 mm bars at $(\pi 8^2/4)/0.481 = 105$ mm centres which could be placed at the top of the slab to help in resisting the transverse flexure; these bars need to be fully anchored across the shear plane or alternatively the bond strength must be equal to $A_{tr}f_{yr}$.

6.5.1.4 **Example 6.4** *Composite L-beams*

The steel component in the composite beam in Figure 6.7 has been placed on the edge of the slab in Figure 6.8 to form a composite L-beam. As the area of slab to the right of the shear plane E-E is much smaller than the area to the left, it can be assumed that $q_E = q_t = 475$ N/mm. As the bond length of the profiled sheeting to the right of E-E is very short '$A_{prof} f_{yp}$' $\rightarrow 0$. Applying Eq. 6.5 gives

$$q_E = 0.66 f_{ct} L_p + 0.8 A_{prof} f_{yp} + 0.8 A_{tr} f_{yr}$$
$$475 \qquad 161 \qquad\qquad 0 \qquad\qquad 0.8A_{tr}f_{yr}$$

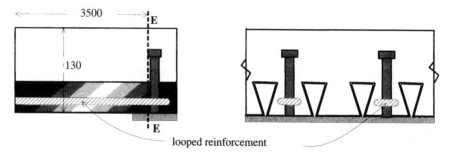

looped reinforcement

Figure 6.8 Composite L-beam

from which $A_{tr} = 0.98$ mm which would require 8 mm bars at 51 mm centres. However, these bars need to be fully anchored across the shear plane E-E which can be achieved by looping the bars around the shear connectors as shown. As each looped bar traverses the shear plane twice, it will only be necessary to space the looped reinforcement at $2 \times$ 51 = 102 mm centres. If this is still too small a spacing, then two bars could be looped around a shear connectors so that the spacing can be increased to 204 mm centres.

6.6 Resistance of shear planes that encompass connectors

6.6.1 *Strength of planes encompassing connectors*

Examples of longitudinal shear planes that encompass the shear connectors are shown in Figure 6.1. Failure can occur along shear planes that encompass individual connectors as in $(L_p)_H$, along shear planes that encompass groups of connectors as in $(L_p)_D$, along shear planes that encompass individual connectors within the group, and across the haunch of a composite beam as in $(L_p)_F$ and $(L_p)_G$. Because these shear planes do not traverse the full depth of the slab as shown in Figure 6.9, the resultant transverse force acting on these shear planes F_{nf} can vary from compression to tension depending on the transverse flexure that is acting on the slab at the time the composite beam is being fully loaded. It is a matter of engineering judgement to determine the magnitude of these transverse forces and the proportion of these transverse forces that act on the shear plane.

The bottom reinforcement A_b in Figure 6.9 crosses the shear plane twice and, hence, the cross-sectional area that resists shear is $2A_b$. In general the slab will be subjected to transverse negative or hogging moments so that the transverse force F_{nf} acting on the shear plane is compressive and as such will enhance the longitudinal shear strength. However, the magnitude of the transverse force depends on the transverse flexure acting on the slab at the time the composite beam is being subjected to its design load and, hence, it is difficult to predict. Furthermore, the portion of this force acting on the shear plane depends on the geometry of the concrete component and in particular the height of the shear plane h_p relative to that of the concrete component h_c and the position of the top reinforcement h_t. The simplest solution is simply to ignore the beneficial effect, that is to assume $F_{nf} = 0$. Inserting these values into Eq. 6.2 and 6.3 gives

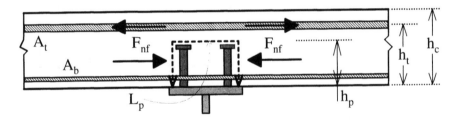

Figure 6.9 Transverse forces on shear planes that encompass shear connectors

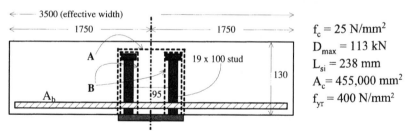

$f_c = 25 \text{ N/mm}^2$
$D_{max} = 113 \text{ kN}$
$L_{si} = 238 \text{ mm}$
$A_c = 455,000 \text{ mm}^2$
$f_{yr} = 400 \text{ N/mm}^2$

Figure 6.10 Shear plane that encompasses the connectors

$$Q_{ch} = 0.66 f_{ct} L_p + 1.6 A_b f_{yr} \qquad (6.8)$$

$$1.6 A_b f_{yr} \geq 0.53 f_{ct} L_p \qquad (6.9)$$

Example 6.5 *Composite T-beam*

Let us now consider the shear flow planes that encompass the shear connectors in Example 6.1 in Section 6.5.1.1. Longitudinal shear failure can either occur around the pair of connectors or around each individual connector as shown in Figure 6.10.

The perimeter length encompassing the pair of connectors in Figure 6.10 is $(L_p)_A =$ $100 + 100 + 95 = 295$ mm. As the area of concrete enclosed by $(L_p)_A$ is much smaller than the cross-section area of the slab A_c, $q_A = q_t = 475$ N/mm. Applying Eq. 6.8 gives $A_b = (475 - 0.66 \times 2 \times 295)/(1.6 \times 400) = 0.134$ mm and Eq. 6.9 gives $A_b > (0.53 \times 2 \times 295)/(1.6 \times 400) = 0.489$ mm and, hence, the lower bound controls the design which will require 8 mm bars at 103 mm centres.

The perimeter length encompassing an individual connector in Figure 6.10 is $(L_p)_B = 219$ mm. Each longitudinal row of connectors transfers a shear flow force of $q/2 = 238$ N/mm. Therefore applying Eq. 6.8 gives $238 = 289 + 640 A_b$, hence only the minimum reinforcement in Eq. 6.9 is needed which gives $A_b > 0.363$ mm. As this transverse reinforcement traverses both shear planes of the individual connectors which are shown as shear planes B in Figure 6.10, this is the total transverse reinforcement required for both shear planes and it is less than that required for the shear plane that encompasses both connectors.

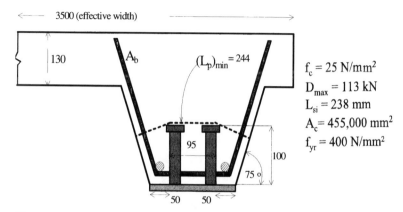

$f_c = 25 \text{ N/mm}^2$
$D_{max} = 113 \text{ kN}$
$L_{si} = 238 \text{ mm}$
$A_c = 455,000 \text{ mm}^2$
$f_{yr} = 400 \text{ N/mm}^2$

Figure 6.11 Haunched beam

6.6.1.2 **Example 6.6** *Composite haunched beam*

If it is assumed that the cross-sectional area of the haunch encompassed by the minimum length of shear plane shown as $(L_p)_{min} = 244$ mm in Figure 6.11 is much smaller than the cross-sectional area of the slab, then the shear flow force on the shear plane is equal to $q_t = 475$ N/mm. Applying Eq. 6.8 gives $A_b = (475 - 66 \times 2 \times 244)/(1.6 \times 400) = 0.239$ mm and Eq. 6.9 gives $A_b = 0.404$ mm. Hence the lower bound controls the design, which requires 8 mm bars at 124 mm centres. These bars need to be fully anchored across this shear plane.

6.7 References

1. Hofbeck, J.A., Ibrahim, I.O. and Mattock, A.H. (1969). 'Shear transfer in reinforced concrete'. ACI Journal, Feb., 119-128.
2. Mattock, A.H. and Hawkins, N.M. (1972). 'Shear transfer in reinforced concrete recent research'. Precast Concrete Institute Journal, March-April, 55-75.

7 Stocky columns

7.1 Introduction

Composite columns are used to resist compressive forces, and most usually act in combination with bending moments. As was noted in Chapter 1, the most commonly used and studied column types are the *encased I-section* shown in Figure 7.1(a), the *concrete-filled steel rectangular hollow section* shown in (b) and the *concrete-filled circular steel tube* shown in (c). Unlike composite beams, the steel component is rarely subjected to tensile forces, and both the concrete and bare steel have to resist compressive stresses.

Short or *stocky* columns fail essentially by squashing, and their strength is governed by the geometric proportions and the material strengths f of the steel and f of the concrete alone. Stocky columns are considered in this chapter. On the other hand, it is common for a column to be quite long or *slender*, for which the strength is governed by overall instability or geometric nonlinear effects. Slender columns are considered in the next chapter.

The column sections shown in Figure 7.1(a) and (b) may be bent about the x-axis by an eccentric compressive load, about the y-axis by an eccentric compressive load or more generally about both axes. In this book, we will consider the case of bending about the major or x-axis of the section, and the methods of analysis may with little modification be extended to bending about the minor or y-axis of the section. The

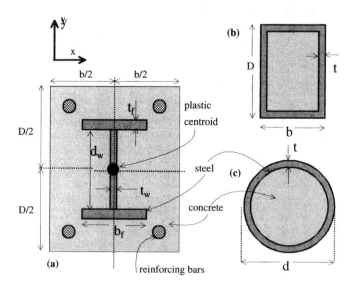

Figure 7.1 Typical symmetrical composite columns

analysis presented in this chapter will assume short-term loading only, so that creep, shrinkage and temperature effects are ignored. This is because we will be concerned with failure of the cross-section, and the combinations of compression and bending actions that cause this failure. The relevant material properties are thus f_y and f_c, and properties modified by long-term effects are not considered.

This chapter will concentrate primarily on encased columns as in Figure 7.1(a) as these are the most commonly encountered form of composite column, although the techniques may be extended easily to concrete-filled rectangular steel sections as in (b), and with a little more difficulty to concrete-filled circular tubes as in (c). The main difference to note is that fully encased steel sections are compact, as they are unable to buckle locally because of the concrete encasement. On the other hand, when the steel is on the outside of the column as in concrete-filled tubes, the geometric provisions of Chapter 2 must be met if the steel is to remain compact and hence achieve its full yield strength.

7.2   Plastic centroid and concentrically loaded column

It is usual for the axial force to be applied at the *plastic centroid* of the section when bending effects are zero, that is for *concentric loading*, or eccentric to the plastic centroid when bending is present.

Consider the non-symmetric column section shown in Figure 7.2(a) that has a cross-sectional area $A_{col} = D \times b$. The cross-sectional area of the steel component is A_s and the remaining area of the column shown shaded is the cross-sectional area of the concrete $A_c = A_{col} - A_s$. Furthermore, the centroid of the steel area A_s is d_s from the top fibre as shown in (d), and the centroid of the concrete area A_c is d_c from the top fibre. The axial strengths of the concrete component P_c and steel component P_s are shown in (b). In order for the concrete component to be uniformly stressed at its ultimate strength, a force F_c of magnitude P_c must be applied at the centroid of the concrete component, that is at a distance d_c from the top fibre as shown in (c) and (d). Similarly, for the steel component to be uniformly stressed at yield, an axial force $F_s = P_s$ must be applied at the centroid of the steel element as shown. The resultant force $P_{sq} = (P_c + P_s)$ is the maximum axial strength of the column and is referred to as the squash load; from equilibrium, the squash load the P_{sq} must act at d_p as shown in (c) and (d) where

$$d_p = \frac{P_c d_c + P_s d_s}{P_c + P_s} \tag{7.1}$$

in which

$$P_c = 0.85 f_c A_c \tag{7.2}$$

$$P_s = f_y A_s \tag{7.3}$$

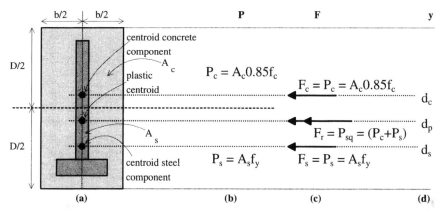

Figure 7.2 Plastic centroid of non -symmetric composite cross-section

The position at d_p is referred to as the plastic centroid. The depths d_c and d_s in Eq. 7.1 are the depths to the geometric centroids of the concrete and steel components respectively, and implicit in Eqs. 7.1 to 7.3 is that the cross-section is rigid plastic, that is the steel and concrete are fully and uniformly stressed.

A more convenient method of determining both the plastic centroid d_p and the squash load P_{sq} is to assume in Eq. 7.2 that A_c is the cross-sectional area of the column A_{col} which equals $b \times D$. This assumption overestimates the cross-sectional area of concrete by an area A_s and, hence, in order to compensate for this additional area of concrete, the yield strength of the steel in Eq. 7.3 is reduced to $f_y - 0.85f_c$. Therefore Eqs. 7.2 and 7.3 become

$$P_C = 0.85 f_c A_{col} \qquad (7.4)$$

$$P_s = \left(f_y - 0.85 f_c\right) A_s \qquad (7.5)$$

However, in many design procedures Eqs. 7.1 to 7.3 are used, with A_c being made equal to A_{col}.

It is worth noting that the familiar elastic centroid is not in general at the same position as the plastic centroid. The position of the elastic centroid d_e below the top fibre of the section in Figure 7.2 is given by

$$d_e = \frac{A_c d_c + n A_s d_s}{A_c + n A_s} \qquad (7.6)$$

where n is the elastic short-term modular ratio E_s/E_c, and d_c and d_s are defined as before. Of course, the denominator in Eq. 7.4 is simply the transformed area.

For doubly symmetric sections as in Figure 7.1(a), the elastic and plastic centroids coincide, and fortunately this is often the case in practice. If the axial compressive load N is applied *eccentrically* at an eccentricity e to the plastic centroid in the direction of the plane of bending, then we will refer the effects of this axial force back to the plastic centroid, which will be assumed to be subjected to an axial compressive force N and a moment Ne in the plane of bending.

7.3 General methods of analysis

7.3.1 Elastic-plastic technique

Figure 7.3 shows an encased I-section member in compression that is also bent about its major or x-axis. We will consider this case, although the extension to bending of an encased I-section about the y-axis and to a concrete-filled rectangular hollow section is simple. The extension to a circular concrete-filled tube is a little more complex, but nevertheless follows the same arguments. The analysis in Figure 7.3 is an elastic-plastic analysis, for which it will be assumed that there is full interaction with only one strain profile. The concrete encases the whole of the I-section in (a) and it is unlikely that all of the steel component will have yielded before the concrete crushes at a strain of 0.003. Hence it will be necessary to use the familiar γ factor method in reinforced concrete design[1] to determine the real neutral axis position n_a below the top fibre shown in (b). The stress distribution in the steel component will be assumed to have the elastic-plastic distribution in (d), while in contrast the concrete component will be assumed to be rigid plastic as in (c), stressed to $0.85f_c$ over a depth γn_a, where[1]

$$\gamma = 0.85 - 0.007\left(f_c - 28\right) \le 0.85 \tag{7.7}$$

where the units of f_c are in N/mm².

With the previous assumptions and for a given neutral axis depth n_a, the resultant force in the concrete F_c, ignoring the presence of the steel component, will be positioned $\gamma n_a/2$ below the top fibre, as in Figure 7.3 (e), and given by

$$F_c = 0.85f_c\gamma n_a b \tag{7.8}$$

where b is the width of the concrete component in (a). Because we know that the curvature in the steel element κ is $0.003/n_a$ as can be seen in Figure 7.3(b), the strain distribution in (b) is uniquely defined, and by invoking the elastic-plastic stress-strain curve 0-C-B in Figure 1.8 for the steel component so too is the stress distribution. As the steel stresses are known, the forces F_1 in the top flange, F_2 in the web and F_3 in the bottom flange may be calculated as shown in (f), where F_3 is negative in this example as the bottom flange is in tension as shown in (d). Because the yield strain of the steel is usually considerably less than the concrete crushing strain (0.003), the steel top flange and some of the web element will often be at yield.

As the forces F_1, F_2 and F_3 act through the centroid of their stress blocks defined over their respective areas, the depths to these forces below the top fibre may be conveniently obtained, and taking the moments of these forces about the top fibre produces the moment M_{top}. Of course, any reference fibre may be used, but the top fibre is often a convenient one, and usually we finally relate the moment and axial force acting in the section to that which occurs at the plastic centroid. From equilibrium, the axial force N_{short} which acts in the cross-section shown in Figure 7.3(g) is clearly the following algebraic sum of the element forces.

$$N_{short} = F_c + F_1 + F_2 + F_3 \tag{7.9}$$

while the moment M_{short} at the plastic centroid is, from simple statics

$$M_{short} = \frac{N_{short}D}{2} - M_{top} \tag{7.10}$$

since the top fibre is positioned D/2 above the plastic centroid.

By varying the position of the neutral axis depth n_a in Figure 7.3, we may determine a locus of points (M_{short}, N_{short}) at which a stocky cross-section fails, as shown in Figure 7.4. This locus is the failure envelope of the section, and generally

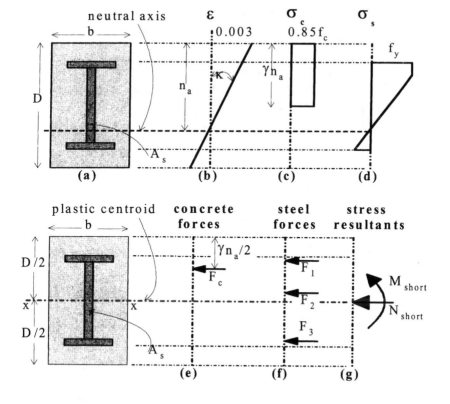

Figure 7.3 Analysis of doubly symmetric section

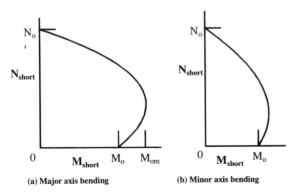

(a) Major axis bending   (b) Minor axis bending

Figure 7.4 Stocky column strength interaction curves

has the shape shown in (a) for major axis bending and in (b) for minor axis bending. If we let $n_a \rightarrow \infty$, then a condition of pure compression will arise and if the area of steel encased by the concrete is ignored in determining the area of the concrete, as was done in the above analysis, then the squash load N_o will be produced as

$$N_O = 0.85bDf_c + A_s f_y \qquad (7.11)$$

Note that the squash load is the same as in the denominator of Eq. 7.1. Clearly we may generate the failure envelope in this fashion by suitable computer programming, but it will be shown in the Section 7.4 how an approximate curve may be produced by assuming rigid plastic behaviour.

The reader familiar with reinforced concrete design will note that the composite strength interaction curves take a similar form to those for reinforced concrete columns. However, the same assumptions are not always applicable to both. The concept of balanced failure in singly or doubly reinforced concrete columns is not the same as that for composite columns. Balanced failure in reinforced concrete columns occurs when the concrete crushes at a strain ε_u of 0.003 and the steel yields in tension simultaneously at a strain of ε_y, and this point on the (M_{short}, N_{short}) curve closely defines the point of maximum moment. In reinforced concrete design, the region below the balanced failure point defines what is called 'tension failure' and above this point defines what is called 'compression failure'. However, this is not the case for composite columns, and in fact the simultaneous failure mode does not appear to be of particular interest for composite columns.

7.3.1.1 **Example 7.1** *Section strength from elastic-plastic assumptions*

Figure 7.5(a) shows an encased I-section bent about the major axis, where the geometric dimensions are in mm and $f_c = 30$ N/mm^2 and $f_y = 300$ N/mm^2. If the neutral axis depth n_a lies at infinity, the concrete stress block is as shown in (b) and the steel stress block is as shown in (c). Since $A_c = 500 \times 900 = 450,000$ mm^2 (ignoring the encased steel) and $A_s = 16,000$ mm^2, from Eq. 7.11 $N_o = 16,275$ kN.

It is worth noting that the true squash load $P_{sq} = P_c + P_s = (450,000 \times 0.85 \times 30)$ + $(16,000 \times (300 - 0.85 \times 30)) = 11,475 + 4,392 = 15,867$ kN which is 2.5% less than N_o, hence the error can be ignored.

The neutral axis is now taken at mid-depth, that is $n_a = 450$ mm, and from Eq. 7.7 $\gamma = 0.836$, so that $\gamma n_a = 376$ mm. The force F_c in Eq. 7.8 is $0.85 \times 30 \times 0.836 \times 450$ $\times 500 = 4797$ kN as shown in (g) which acts $376/2 = 188$ mm from the top fibre as shown in (h). It is worth re-emphasizing that in Eq. 7.8 the area of the steel within the concrete component has been ignored.

From Figure 7.5(d), the strain on the inside face of the flange is $((450 - 100)/450)$ $\times 0.003 = 0.00233 > \varepsilon_y = 300/200 \times 10^3 = 0.0015$, so that all of the top flange has yielded as in (f). Hence from symmetry $F_{s1} = -F_{s4} = 300 \times 15 \times 300$ N $= 1350$ kN as shown in (g). The depth in the steel below the top fibre to first yield is $450 - (0.0015/\,0.003) \times 450 = 225$ mm, so that $450 - 225 = 225$ mm of the web above the

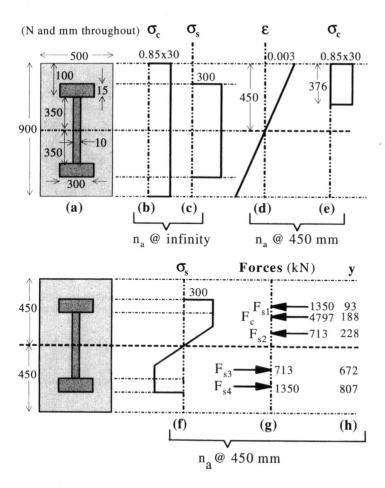

Figure 7.5 Example 7.1: Elastic-plastic axial strength

neutral axis is elastic and $350 - 225 = 125$ mm of the web above the neutral axis is at yield. The total force in the web above the neutral axis F_{s2} is then $(125 + 225/2) \times 10 \times 300$ N $= 713$ kN, and the height at which it acts above the neutral axis is $[(125 \times (225 + 125/2)) + (225^2/3)] / (125 + 225/2) = 222$ mm or 228 mm below the top fibre. The force in the web below the neutral axis F_{s3} is clearly -713 kN and acts $450 + 222 = 672$ mm below the top fibre.

The forces in Figure 7.5 can be summed to produce $N_{short} = 4797$ kN, and the moment of these forces about the top fibre can be summed to produce $M_{top} = 4797 \times 188 + 1350 \times 93 + 713 \times 228 - 713 \times 672 - 1350 \times 807$ kNmm $= -379$ kNm. Hence from Eq. 7.10, $M_{short} = -379 + 4797 \times 900 \times 10^{-3} / 2 = 1780$ kNm. This point $(1780, 4797)$ may be plotted on the strength interaction envelope, and corresponds to an eccentricity e_1 of $1780 \times 10^3/4797 = 371$ mm. The above analysis may clearly be programmed on a spreadsheet, giving values of (M_{short}, N_{short}) as a function of n_a that may be used to generate a locus corresponding to the interaction diagram by varying n_a from zero to a large number. When n_a is small, the cross-section will be subjected to tension, and this condition is generally ignored in deriving the strength interaction diagram.

7.3.2 Rational non-linear analysis

The method described in Section 7.3.1 above forms the basis of a more rational technique for determining the strength interaction diagram. This method has been confined to research analyses, in which a particular axial force N_{short} is assumed in the range $0 \leq N_{short} \leq N_0$. For this particular value of $N_{short} = (N_{short})_1$ a small curvature κ_1 is chosen. The strains corresponding to this value of κ_1 are still not yet defined, as we do not yet know the position of the neutral axis n_a. This is determined by varying n_a, with each value of n_a and the assumed value of κ_1 uniquely describing the strain distribution in the cross-section, as in Figure 7.3(b). As the strains are known at each point in the cross-section, the accurate stress-strain relationships described in Chapter 1 may be invoked to determine the stress in the concrete, steel or reinforcement at each point in the cross-section. These stresses are then integrated numerically over the cross-section to determine the axial force N that equates to the combination of κ_1 and n_a. If $N \neq (N_{short})_1$, the neutral axis depth n_a is adjusted progressively until N converges to $(N_{short})_1$ with an acceptable tolerance. At this stage, for the predetermined value of $(N_{short})_1$ are a curvature κ_1, and the strain (and stress) distribution throughout the cross-section is defined uniquely.

The next step is to integrate the first moments of the stresses over the cross-section in order to determine the moment M_1 at the plastic centroid which gives point 1 in Figure 7.6. The curvature κ_1 is then increased to κ_2, with the iterative scheme again being used to generate the moment M_2 for the chosen value of $(N_{short})_1$ and a given curvature κ_2 and hence point 2 in Figure 7.6. In this way, we may generate the moment versus curvature curve for a given $(N_{short})_1$ as shown in Figure 7.6. The curve shown is this figure demonstrates that, after a peak moment is attained, the curve then softens owing to the presence of the descending branch of the stress-strain curve for concrete shown schematically in Figure 1.10. Fortunately, there are empirical equations that express the stress-strain curves for the concrete, steel and

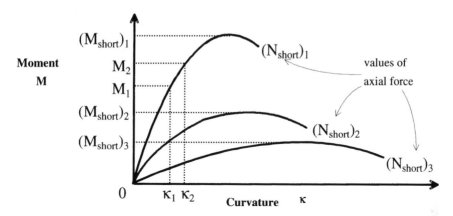

Figure 7.6 Moment-curvature response for a given axial force

reinforcement, so that generation of the moment-curvature curve may be performed by routine computer programming. The maximum moment that is attained in Figure 7.6 is identified as $(M_{short})_1$, so that we now have one point $(N_{short})_1$, $(M_{short})_1)$ on the strength interaction diagram in Figure 7.4.

The entire interaction diagram for a stocky column may be generated in this fashion, by specifying a value of $(N_{short})_n$, calculating the moment-curvature relationship corresponding to this load and identifying the peak moment $(M_{short})_n$. The process is rather involved mathematically, as for each assumed $(N_{short})_n$, an iteration must be performed at each curvature to determine the position of the neutral axis. However, the method is well-suited to obtaining rapid solutions on modern computers, and has the main advantage that the actual stress-strain curve for the concrete is used, so we do not have to rely on the approximation of a rectangular stress block stressed to $0.85f_c$ over a depth γn_a. Of course, this rational approach may be employed with little modification to short concrete-filled steel tubes.

7.4 Rigid plastic analysis

7.4.1 *General*

In the ingenious method developed by Roik and Bergmann[2], the interaction diagram AECDB shown in Figure 7.7 is approximated by the polygon ACDB shown. The approach assumes that the cross-section is doubly symmetric about the axis of bending, which is usually the case, and is based on rigid plastic principles described in Chapters 1 and 4, so that the steel or concrete is either fully yielded or not stressed at all. The procedure will be illustrated for the general cross-section shown in Figure 7.8(a), which is symmetrical about the axis D which is the position of the plastic centroid. Because of the symmetry of the cross-section about D; the plastic centroid is also the position of the elastic centroid. The section is comprised of the steel component, which is assumed to be fully yielded in compression and tension at f, and the concrete component which is assumed to be fully yielded in compression at

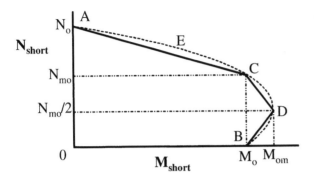

Figure 7.7 Ultimate strength failure envelope

0.85f , but which has no tensile strength. The following definition of each of the points in Figure 7.7 will be illustrated by means of an example. Of course, the point A in Figure 7.7 is obtained from Eq. 7.11.

Point 'B'

7.4.2.1   Pure bending analysis

Point B in Figure 7.7 corresponds to the pure flexural capacity. Consider the distribution of stress shown in Figure 7.8(b) where the stress above the neutral axis N-A is compressive. Below the neutral axis, only the steel component is stressed in tension, and this neutral axis is positioned h_n above the plastic centroid. For convenience, the section is divided into three regions shown in (a), namely region 2 which lies h_n on either side of the plastic centroid, region 1 which is further than h_n above the plastic centroid and region 2 which is further than h_n below the plastic centroid. The resultant forces in the steel and concrete components are shown in (c), where F_{c1} is the resultant force in the concrete in region 1 and acts at the centroid of the concrete region, while F_{s1} is the resultant force in the steel component in this region and acts through the centroid of the steel in region 1, and so on. Taking moments of the forces in (c) produces the pure flexural moment capacity M_o shown in (d).

The force F_{s2} in region 2 in Figure 7.8(c) must act through the plastic centroid because of the symmetry about D in Figure 7.8 that we have assumed. This symmetry also dictates that the magnitude of the force $F_{s1} = F_{s3}$. As the section is in pure flexure, the net resultant of the axial forces is zero, that is

$$F_{c1} + F_{s1} = F_{s2} + F_{s3}$$
(7.12)

and as it has been shown that $F_{s1} = F_{s3}$ then $F_{c1} = F_{s2}$. Furthermore as the area of concrete in regions 1 and 3 in (a) are the same, then the compressive strength of the concrete in region 3, that is F_{c3}, is the same as that in region 1, then $F_{c3} = F_{c1}$. Hence

$$F_{c1} = F_{c3} = F_{s2}$$
(7.13)

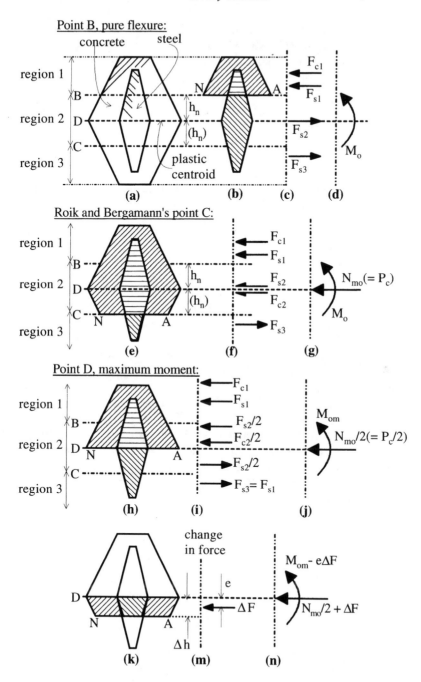

Point B, pure flexure:

Point D, maximum moment:

Roik and Bergamann's point C:

Figure 7.8 Column interaction points for general cross-section

7.4.2.2   **Example 7.2** *Roik and Bergann's method – Point B*

Consider the cross-section shown in Figure 7.5(a) that was considered in Example 7.1. If it is assumed that the neutral axis lies in the web of the steel component and using the notation in Figure 7.8, then $F_{s2} = 2h_n \times 10 \times 300 = 6000h_n$ N, while the force in the concrete component $F_{c1} = (450 - h_n) \times 500 \times 0.85 \times 30 = 12750 \times (450 - h_n)$. Since $F_{c1} = F_{s2}$ from Eq. 7.13, $h_n = 306$ mm which indeed is in the web element, and $F_{c1} = 0.85 \times 30 \times 500 \times (450 - 306) = 1836$ kN. The force F_{s1} in region 1 is then $300 \times (15 \times 300 + (350 - 306) \times 10)$ N $= 1482$ kN and is positioned $(15 \times 300 \times 358 + 10 \times (350 - 306)^2/2) \times 300/1482 \times 10^3 = 328$ mm above the plastic centroid. Moreover, the concrete force F_{c1} is positioned $(450 - 306)/2 + 306 = 378$ mm above the plastic centroid. Hence by taking the moment of the forces F_{c1}, F_{s1} and F_{s3} about the plastic centroid produces $M_o = 1836 \times 378 + 1482 \times 328 + 1482 \times 328$ kNmm $= 1666$ kNm.

7.4.3 *Point 'C'*

7.4.3.1   Analysis

Point C in Figure 7.7 lies where the moment capacity about the plastic centroid is the same as the pure flexural capacity M_o, but where there is a resultant axial force N_{mo}. The analysis of this section is shown in Figures 7.8(e) and (f). Roik and Bergmann showed[1,2] that the neutral axis N-A in (e) lies h_n below the plastic centroid, where h_n was determined as in Section 7.4.2.1 for point B. This can be confirmed by comparing the forces in (c) with those in (f). The only difference between these force distributions is F_{c2} in (f) and as this acts through the plastic centroid, as the steel element is symmetrical about the plastic centroid, F_{c2} does not contribute to the moment and, hence, the moments at Points B and C are the same.

The resultant force in Figure 7.8(c) is zero as we are dealing with pure flexure. As the moment M_o is the same in (c) and (f), any changes between these resultant forces due to the movement of the neutral axis from the level in (b) to that in (e) is equal to the increase in the axial force N_{om}. Comparing (f) with (c), it can be seen that the algebraic change in the steel force is $2F_{s2}$ and the change in the concrete force is F_{c2}, therefore the change in force

$$N_{mo} = 2F_{s2} + F_{c2} \tag{7.14}$$

and substituting Eq. 7.13 gives $N_{mo} = F_{c1} + F_{c3} + F_{c2}$, that is

$$N_{mo} = P_c \tag{7.15}$$

where P_c is the axial compressive strength of the concrete component.

7.4.3.2   **Example 7.3** *Roik and Bergmann's method – Point C*

Consider the cross-section in Figure 7.5(a). We saw in Example 7.2 that $F_{c1} = 1836$ kN and lies $(450 - 306)/2 + 306 = 378$ mm above the plastic centroid, $F_{s1} = 1482$ kN and lies 328 mm above the plastic centroid, and $F_{s3} = 1482$ kN and lies 328 mm below the plastic centroid. Moreover, h_n again equals 306 mm, but this time is below the plastic centroid.

Clearly, $F_{c2} = 2 \times 306 \times 500 \times 0.85 \times 30 \, N = 7803 \, kN$. The axial compressive force is N_{mo} = 1836 + 7803 = 9,639 kN which is equal to the axial compressive strength of the concrete P_c and taking moments about the plastic centroid produces M_o as in Example 7.2.

7.4.4 *Point 'D'*

7.4.4.1   Analysis

Point D in Figure 7.7 corresponds to the point of maximum moment. Roik and Bergmann showed that the neutral axis for this point passes through the plastic centroid as shown in Figure 7.8(h). The proof is illustrated in (k) to (n). If the neutral axis is dropped a fraction Δh then the additional axial compressive force ΔF reduces the moment M_{om} in (j) to $M_{om} - e\Delta F$ shown in (n). Furthermore, if the neutral axis is raised then the change in the axial force ΔF is now tensile and acting above the original level of the neutral axis so reducing the moment to that shown in (n). Hence the maximum moment occurs when the neutral axis passes through the centroid.

The summation of the forces in Figure 7.8(i) gives

$$N_{short} = F_{c1} + \frac{F_{c2}}{2} = \frac{N_{mo}}{2} = \frac{P_c}{2} \qquad (7.16)$$

that is half the compressive strength of the concrete component. The concrete forces $F_{c2}/2$ and F_{c1} in (i) lie above the neutral axis, while the compressive and tensile steel forces $F_{s2}/2$ lie at the centroid of the steel in region 2 above and below the neutral axis respectively. It is clear that the moment contribution of the steel component is equal to its plastic moment capacity.

7.4.4.2   **Example 7.4** *Roik and Bergmann's points D and A*

We have already established in Examples 7.2 and 7.3 that F_{c2} = 7803 kN, F_{c1} = 1836 kN (378 mm) and F_{s1} (328 mm) = F_{s3} (−328 mm) = 1482 kN, where the distance in brackets is the height of the force above the plastic centroid, which is also the position of the neutral axis. Clearly the force $F_{c2}/2$ lies 306/2 = 153 mm above the neutral axis, and because only the web element of the steel component lies in region 2, $F_{s2}/2 = 306 \times 10 \times 300 \, N = 918 \, kN$ and lies 306/2 = 156 mm above the plastic centroid in compression and 156 mm below the plastic centroid in tension. From Example 7.3, the axial force N_{short} = 9639/2 = 4820 kN and using Example 7.2, the moment about the plastic centroid is M_{om} = 1666 + 2×918×156×10⁻³ + (7803/2) × 153×10⁻³ = 2549 kNm.

Finally, Point A in Figure 7.7 is N_o = 16,275 kN from Example 7.1.

Points A, B, C and D are shown plotted in Figure 7.9 for Examples 7.2 to 7.4. It is worth noting that in Example 1, at a load eccentricity of 371 mm, the axial strength was 4797 kN and the bending strength was 1780 kNm. A line corresponding to this eccentricity is shown in Figure 7.9, where the point on the rigid plastic failure envelope is (2280 kNm, 6100 kN). It can thus be seen that the rigid plastic assumption of Roik and Bergmann overestimates the bending and the axial compressive capacities by 28%. This degree of unconservatism can be tolerated, as it can be argued that strength

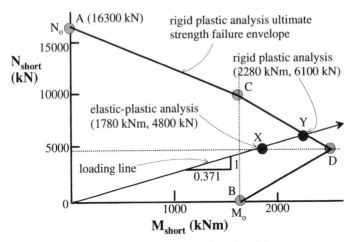

Figure 7.9 Failure envelopes from Examples 7.1 to 7.4

reduction factors in national codes of practice, in conjunction with the benign strain hardening effect of the steel component, will compensate for the unconservative assumptions made in the rigid plastic assumption.

7.4.5 *Allowance for shear*

The rigid plastic analysis must be modified for the effects of shear, in the same way as in Chapter 4. If the composite column is loaded with unequal end eccentricities, it will experience a moment gradient and hence a shear force V. This shear force is assumed to be resisted by the web element of the steel component.

Because of this, the web yield stress in the method of Roik and Bergmann must be reduced below its strength f_y. It was shown in Section 4.2.5 that an appropriate reduced yield strength is f_{fy} given by Eq. 4.2, where the shear stress in the web of depth d_w and thickness t_w is taken as

$$\tau_w = \frac{V}{d_w t_w} \tag{7.17}$$

Equations 4.2 and 7.17 are based on the von Mises yield criterion (Chapter 1), and on the assumption that the shear yield stress τ_w in the web element of the steel member is uniform.

7.5 References

1. Oehlers, D.J. and Bradford, M.A. (1995). Composite Steel and Concrete Structural Members: Fundamental Behaviour. Pergamon Press, Oxford.
2. Roik, K. And Bergmann, R. (1989). 'Eurocode 4: composite columns'. Report EC4/6/89, University of Bochum, June.

8 Slender columns

8.1 Introduction

In Chapter 7, we introduced the concept of a composite column that is required to resist axial compression as well as bending actions. The behaviour of the column was restricted to stocky columns whose strength was attributable entirely to material failure of the cross-section. Composite columns in practice are rarely stocky, but are *slender*, in that their length is much larger than the cross-sectional proportions. This chapter will again consider composite columns which are subjected to compressive and bending actions, but the behaviour is inherently more complex as the column slenderness contributes to the so-called second order effects. These second order effects must be considered in the design of slender composite columns. The fundamental procedures will first be developed for steel columns in Sections 8.2 and 8.3 which are then applied to composite columns in Section 8.4.

8.2 Elastic columns

8.2.1 *Concentric loading*

8.2.1.1 First yield approach

Flexural or Euler buckling is based on elastic principles, and forms the basis for the design of concentrically loaded steel columns. We will consider the analysis of this type of buckling here, as the results are used for the design of composite columns described in Section 8.4. Euler buckling of a simply supported and initially straight steel column takes place when the column moves to an adjacent equilibrium position at a load N_E given by[1]

$$N_E = \frac{\pi^2 E_s I}{L^2} \tag{8.1}$$

where L is the length of the column and I is its second moment of area about the weaker principal axis. The shape of the buckle follows a sine curve, but the magnitude of the buckle is indeterminate. The concept of Euler buckling is used to some extent in reinforced concrete design, but because of the material nonlinearities in the variation of E_c and particularly due to cracking, a concrete column will not 'buckle' at a load given by Eq. 8.1. This is also true of a composite column, as the concrete component also behaves nonlinearly with respect to its material properties. However, it is worth reiterating that the most advanced analysis techniques for composite columns, which will be treated in this chapter, are in fact based on the concept of Euler buckling.

If a concentrically loaded (steel) column was 'perfect' in the sense that it was initially straight, concentrically loaded and had the stress-strain curve given in Figure 3.1, it would fail by Euler buckling at N_E for lengths greater than $\pi\sqrt{(E_s I/N_{sq})}$ and

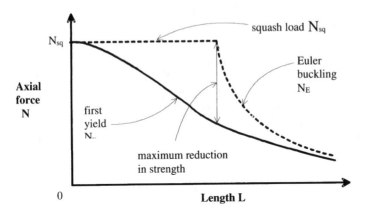

Figure 8.1 Failure of an elastic steel column

squash at lengths less than this, where $N_{sq} = A_s f_y$ is the squash load of the steel cross-section. This failure load is shown in Figure 8.1.

Steel columns are never 'perfect'. We will now drop the condition that the column is initially straight, and suppose that it has an initial imperfection u_o that varies sinusoidally according to

$$u_o = \delta_o \sin \frac{\pi z}{L} \qquad (8.2)$$

where z is measured from the end of the column along the length, and δ_o is the maximum value of the initial out of straightness as shown in Figure 8.2. For this crooked column, the member deflects increasingly by an additional amount δ at midspan as the axial load N is increased, given by[1]

$$\frac{\delta}{\delta_o} = \frac{N/N_E}{1 - N/N_E} \qquad (8.3)$$

It is noteworthy that when N = 0 in Eq. 8.3 then $\delta = 0$ as expected, and that as $N \to N_E$ then the column deflection increases towards infinity. This high column curvature as N increases leads to increased stresses in the column, and forms the basis of the failure theory outlined below.

The bent column under an axial load N is subjected to a moment at midspan of $N(\delta + \delta_o)$ as can be seen in Figure 8.2 which, on the concave side of the bent column, produces a compressive stress of $N(\delta + \delta_o)/Z$, in which Z is the section modulus. The axial compression also produces a compressive stress N/A_s. If the column is deemed to fail when the sum of these two compressive stresses reaches the yield stress of the steel f_y then

$$f_y = \frac{N}{A_s} + \frac{N(\delta + \delta_o)}{Z} \qquad (8.4)$$

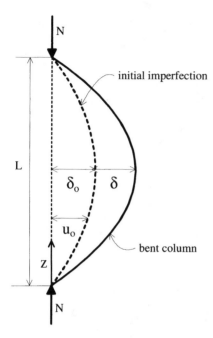

initial imperfection

L

δ_o δ

u_o

Z

bent column

N

Figure 8.2 Initial imperfections in a column

The load N_y at which first yield occurs may be obtained by substituting Eq. 8.3 into Eq. 8.4. After some manipulation, this load may be written as

$$N_y = N_{sq}\left\{\xi - \sqrt{\xi^2 - \frac{N_E}{N_{sq}}}\right\}$$ (8.5)

in which

$$\xi = \frac{1 + (1 + \eta)\dfrac{N_E}{N_{sq}}}{2}$$ (8.6)

and η is called the *imperfection parameter*, which in this derivation is given by

$$\eta = \frac{\delta_o D}{2r^2}$$ (8.7)

where D is the width of the member (assumed to be rectangular) transverse to the axis of buckling and $r = \sqrt{(I/A_s)}$ is its radius of gyration.

8.2.1.2 **Example 8.1** *Strength of a simply supported concentrically loaded steel bar*

Consider the case of a simply supported rectangular steel bar of dimensions $200\,\text{mm} \times 20\,\text{mm}$ and yield stress $f_y = 300\,\text{N/mm}^2$. Its properties are $A_s = 200 \times 20 = 4000$

mm^2 and I = $20^3 \times 200/12 = 133 \times 10^3$ mm^4 and so r = $\sqrt{(133 \times 10^3/4000)} = 5.77$ mm. The squash load is N$_{sq}$ = 4000 × 300 N = 1200 kN, and let us assume further that the length is L = 468 mm, so that its Euler load from Eq. 8.1 is N$_E$ = $\pi^2 \times 200 \times 133 \times 10^3/468^2$ = 1200 kN which is the same as the squash load. Hence if the column was 'perfect', Euler buckling and yielding would occur simultaneously at a load of 1200 kN.

Suppose now that the column has a maximum out of straightness δ_0 = 5 mm. From Eq. 8.7, η = 5 × 20/(2 × 5.77^2) = 1.50, from Eq. 8.6 ξ = (1 + (1 + 1.50) × 1200/1200)/2 = 1.75 and so from Eq. 8.5 N$_y$ = (1.75 − $\sqrt{(1.75^2 - 1200/1200)}$) × 1200 = 377 kN. The presence of the geometric imperfection that renders the column crooked thus reduces the strength of the column by 377/1200 or 69% below that assuming elastic buckling.

The load to cause first yield of a crooked column N$_y$ is shown schematically in Figure 8.1. As L → 0 the strength approaches the squash load N$_{sq}$ of the column, while as L → ∞ the effects of yielding become negligible and the strength approaches the Euler load N$_E$. The greatest reduction in strength below either N$_{sq}$ or N$_E$ actually occurs at the point where N$_{sq}$ = N$_E$ which was considered in Example 8.1.

8.2.1.3 Column curves

Equation 8.5 forms the basis of column strength curves for steel columns that are given in many national standards. In the derivation presented in Section 8.2.1.1, it was assumed that the column was rectangular, and that moreover the magnitude of the initial geometric imperfection δ_0 was known. Both of these effects appear in the imperfection parameter η.

In a real (steel) column, the cross-section is rarely rectangular and the magnitude of the initial out-of-straightness is also unknown. In addition, residual stresses that are formed during the manufacture of the column are present, but are not easily quantifiable. The column strengths given in national standards are therefore based on Eq. 8.5, but they have been calibrated against test results and also against advanced numerical solutions that incorporate a variety of cross-sections, geometric imperfections and residual stresses. By doing this, the imperfection parameter η may be determined empirically as a function of the slenderness ratio λ = L/r (where r is the radius of gyration about the axis of buckling) and also as a function of the type of cross-section, which categorizes its residual stress pattern. When this is simplified, the strength of steel columns are categorized by their slenderness and their cross-sectional type. For example, in the Eurocode[2], Eq. 8.5 is written as

$$N_{col} = \chi N_{sq}$$

(8.8)

where N$_{col}$ is the strength of the steel column, N$_{sq}$ is its squash load and the slenderness parameter χ is a function of the slenderness ratio λ = L/r and also of the empirically determined imperfection parameter η.

If we write the term N$_E$/N$_{sq}$ that occurs in Eqs. 8.5 and 8.6 as

$$\frac{N_E}{N_{sq}} = \frac{\pi^2 E_s I}{L^2 A_s f_y} = \frac{1}{\lambda^2} \frac{\pi^2 E_s}{f_y}$$

(8.9)

then clearly N_E/N_{sq} is inversely proportional to the slenderness ratio squared. The result of this is that the column curve concept expresses the slenderness parameter χ as a function of the slenderness ratio λ. In the Eurocode[2], four column curves are given as illustrated in Figure 8.3, and these curves are denoted a,b,c and d. Column 'a' is used for sections that are free from residual stresses, which calibration of Eq. 8.5 with tests show have a very low value of η and hence have the highest strength. On the other hand, the column curve labelled 'd' is used for sections with thick plates, which experimental calibration shows that η is quite large, and so the corresponding strength is lower. The curves labelled 'b' and 'c' apply to cross-sections whose residual stresses and out-of straightness due to fabrication have been found to be between those 'a' and 'd'. It is worth noting that the British steel standard also presents four column curves, while the Australian standard presents five.

8.2.1.4   Effective lengths

The length of the column L in the derivation of the column curve was taken as that between simple supports as shown in Figure 8.4(a). The elastic buckling load N_{crit} of a perfect column depends on what is termed the *effective length* of the column. For example, if the column of length L was not simply supported at each end, but in fact was fully built-in at each end as in (d), then its buckling load can be shown to be $4\pi^2 EI/L^2$ or $4N_E$, that is the buckling load of the portion of the column or idealized column shown as L_e that resembles the simply supported Euler column. It is most usual to express the end conditions of a column by referring to its effective length L_e, so that

$$N_{crit} = \frac{\pi^2 EI}{L_e^2} \qquad (8.10)$$

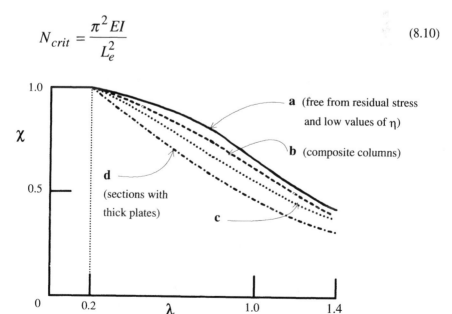

Figure 8.3 Illustration of column curves used in the Eurocode (not to scale)

which merely replaces the length L in the Euler formula by the effective length L_e. Further, it is common to express the effective length of a column of length L in terms of its effective length factor k_e by

$$L_e = k_e L \qquad (8.11)$$

Further examples of effective length factors for idealized conditions of column restraint are given in Figure 8.4.

Generally, a column in a building does not have the idealized end restraints shown in Figure 8.4, but is part of a rigid frame so that the end condition is intermediate between fully built-in at both ends and simply supported at both ends. In addition, the effective length is influenced by whether or not the frame of which the column is a part is free to sway. The determination of the effective length factor k_e in these cases is beyond the scope of this book, but is explained for elastic columns in Ref. 1. Guidance for determining the effective length factor is generally given in chart or nomogram form in most national steel (and concrete) standards, in which the effective length is a function of the stiffness of restraining beams and columns at each end of the column under consideration.

The concept of the effective length is important, as the slenderness ratio is given more generally, not as L/r, but as

$$\lambda = \frac{L_e}{r} \qquad (8.12)$$

This slenderness ratio is again used in the column curves, in which the column effective length L_e merely replaces the simply supported length L.

Example 8.2 *Strength of a restrained concentrically loaded steel bar*
Let us consider the steel bar that was analysed in Example 8.1, but this time it shall be assumed that the bar is fixed rigidly at both ends. As in Example 8.1, we shall make recourse to first principles and not to any national design standard.

From Figure 8.4(d), the effective length factor is $k_e = 0.5$, so that $L_e = 0.5 \times 468 = 234$ mm and the elastic buckling load is $N_{crit} = \pi^2 \times 200 \times 133 \times 10^3/234^2 = 4795$ kN. Again assume $\delta_o = 5$ mm, so that $\eta = 1.50$ as before. The parameter $\xi = ((1 + (1 + 1.5) \times 4795/1200)/2 = 5.49$, and so from Eq. 8.5 $N_y = (5.49 - \sqrt{(5.49^2 - 4795/1200)}) \times 1200 = 452$ kN. It can thus be seen by comparison with Example 8.1 that increasing the restraint of the column from simply supported ends to built-in ends increases the elastic buckling load by a factor of four, but the effects of yielding result in an increase in failure strength of 452/377 or 20%. Nevertheless, this 20% increase cannot be ignored in design, and so the effective length factor k_e should be used. This is particularly so when $k_e > 1$ (as in columns in frames that are free to sway), as the strength of the slender column derived assuming $k_e = 1$ (Eq. 8.1) will be unconservative.

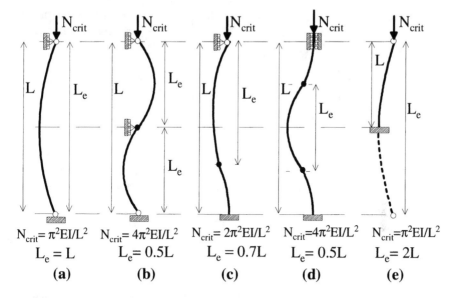

$N_{crit}= \pi^2EI/L^2$ $N_{crit}= 4\pi^2EI/L^2$ $N_{crit}= 2\pi^2EI/L^2$ $N_{crit}=4\pi^2EI/L^2$ $N_{crit}=\pi^2EI/L^2$

$L_e = L$ $L_e= 0.5L$ $L_e = 0.7L$ $L_e= 0.5L$ $L_e= 2L$

(a) **(b)** **(c)** **(d)** **(e)**

Figure 8.4 Effective length factors for idealized end restraints

8.3 End moments

8.3.1 *Secondary effects*

Let us consider a simply supported column that is subjected to end moments M_{m1} and M_{m2} that bend the column into single curvature as shown in Figure 8.5(a). In concrete and composite terminology, this is still referred to as a *column*, while in steel design it is referred to, perhaps more correctly, as a *beam-column*. Although we are discussing steel members here, we will however refer to the member as a column.

The effect of the end moments M_{m1} and M_{m2} is to bend the column into single curvature, as shown in Figure 8.5. This bending curvature gives rise to additional deflections u, and so an additional moment of Nu is generated along the member. The total moment at any cross-section where the deflection is u_o is $M_m + Nu_o$, this moment causes a curvature $(M_m + Nu_o)/EI$, and so we can calculate a new deflection u_1, a new moment $(M_m + Nu_1)/EI$, a revised deflection u_2 and so on. Finally, the bending moment converges to $M_m + Nu$. As shown in Figure 8.5, the constant moment M_m is referred to as the *primary moment*, while the additional moment Nu is referred to as the *secondary moment*. The maximum value of the second order moment, the sum of the primary and secondary moments, is denoted M_{max}, and it is this moment capacity at a given axial compression N that must be established.

Fortunately for elastic analysis a closed form solution for the maximum value M_{max} exists, and a derivation is given in Ref. 1. However, material nonlinearities and cracking in composite columns preclude expressing M_{max} in a closed form, and so approximations have to be made. If for the present we ignore these material nonlinearities that are germane to the concrete component of a composite column and consider again a steel column, in lieu of the complex closed form solution for M_{max} a suitable linear interaction between the moment M_{max} and axial compression N to cause failure of a steel column is

$$\frac{M_{max}}{M_p} = 1 - \frac{N}{N_{col}} \qquad (8.13)$$

where M_p is the plastic moment for bending about the weaker axis.

In Eq. 8.13, the moment M_{max} is the maximum second order moment obtained from the closed form solution. As this closed form solution is complex even for steel, and indeed does not exist for composite columns, the following approximation is used.

$$\frac{M_{max}}{M_{m1}} = \frac{c_m}{1 - \dfrac{N}{N_{crit}}} \qquad (8.14)$$

where N is the axial force, N_{crit} is given by Eq. 8.10, and the factor c_m is intended to take account of the different moments M_{m1} and M_{m2} applied at the ends of the member shown in Figure 8.5. If M_{m1} is chosen as the larger moment and M_{m2} is chosen as the smaller, then the moment gradient β is given by

$$\beta = \frac{M_{m2}}{M_{m1}} \qquad (8.15)$$

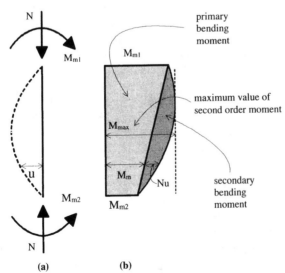

Figure 8.5 Primary and secondary moments in compression members

where β is negative if the end moments bend the column into single curvature and is positive if the end moments bend the column into reverse curvature. Using this definition of the moment gradient, a commonly used expression for c_m is

$$c_m = 0.66 - 0.44\beta \geq 0.44 \qquad (8.16)$$

Hence in Eq. 8.14, the coefficient c_m accounts for the benign effect of the moment gradient, while the denominator, which is less than unity, accounts for the amplification of the moment due to nonlinear effects.

8.3.1.1 Example 8.3 *Moment capacity of a steel square column section*

Let is again consider the column of Example 8.2, but this time the column is subjected to an axial load $N = 300$ kN which has a moment applied at its top and none at its bottom. The moment gradient is thus $\beta = 0$ and so from Eq. 8.16, $c_m = 0.66$. We will use the column strength as determined from first principles rather than that implicit in codes of practice, so that $N_{col} = N_y = 452$ kN. The rigid plastic moment M_p for bending about the weaker axis is $300 \times 20^2 \times 200/4$ Nmm = 6.00 kNm. The second order moment to cause failure from Eq. 8.13 is $M_{max} = (1 - 300/452) \times 6 = 2.02$ kNm. Substituting this value of M_{max} into Eq. 8.14 produces $M_{ml} = 2.02 \times (1 - 300/452)/0.66 = 1.03$ kNm. This end moment would correspond to a load eccentricity at the top of the column of $1030/300 = 3.4$ mm.

8.3.2 *Graphical interpretation*

The provision of Eq. 8.13 may be explained conveniently in graphical form. This is introduced here for steel columns, because it forms the basis of the Eurocode[2] approach for composite columns described in Section 8.4.

Figure 8.6 shows the interaction between compression and bending for a steel column that is similar to Figure 7.4. The moments are plotted dimensionally as $m = M_{short}/M_p$ and the loads as $n = N_{short}/N_{sq}$, where M_p is the plastic moment and N_{sq} is its squash load. The *section failure envelope* is taken in most national codes as the straight line $m = 1 - n$, as shown.

Now consider the point at A in Figure 8.6 where $n_c = N_{col}/N_{sq}$ and $m = 0$, this point corresponds to the pure compressive load to cause failure of the column when loaded concentrically, as would be determined from the relevant column curve. In terms of Eq. 8.8, n_c is identical to χ. If section failure, that is failure of the material such as that given in Figure 7.7, was the relevant limit state, then the column loaded to n_c could support the moment m_{A-B} drawn in Figure 8.6. Of course, the limit state is failure by buckling, so that the column *cannot* resist any of the moment m_{A-B}, as it buckles at n_c without any moment being applied. At the origin when there is no axial force, the column can of course resist the full plastic moment $m = 1$ at point C.

Let us now draw a line (which we will assume to be straight) between B and 0 in Figure 8.6. For any non-dimensional load $n_l = N_l/N_{sq}$ between n_c and 0 as shown in Figure 8.6, the triangular region A-B-0 represents the moment that cannot be supported owing to slenderness or buckling effects. Hence, for a load N_1 that is non-dimensionalized as n_1 as shown in the figure, the moment m_{D-E} cannot be resisted by the column, but the reserve of capacity m_{E-F} *can* be resisted by the column and it is this reserve capacity m_{E-F} that is available to resist M_{max} in Figure 8.5. By noting that the failure envelope for the interaction between n and m is a straight line, and that the region A-B-0 is a triangle, it can be shown from similar triangles that m_{E-F} = $1 - n_1/n_c$. This construction of the reserve capacity m_{E-F} leads to an equation for the capacity of a column under a load $N \le N_{col}$ that is identical to Eq. 8.13. The moment M_{res}/M_p, where M_{res} is the reserve moment capacity, represented by E-F must, of course, be transformed back to end moments in accordance with Eqs. 8.14 to 8.16 as M_{res} is the second order moment M_{max} in Eq. 8.14.

8.4 Moment capacity of slender composite columns

8.4.1 *Concentrically loaded columns*

8.4.1.1 Critical or buckling load

The most widely accepted approach for determining accurately the bending capacity of a composite column is that used in the Eurocode[2], and follows very closely the same arguments presented in Section 8.3 for steel columns. In the previous section, the second order moment was quantified fairly accurately by an approximation based on a knowledge of the buckling strength N_{col} of a steel column. The same rationale is used in composite columns, in that Eq. 8.14 is used to magnify the maximum end moment M_{ml} that has been determined from a first order linear analysis. However, in the Eurocode approach, the 'buckling' load N_{crit} is written as

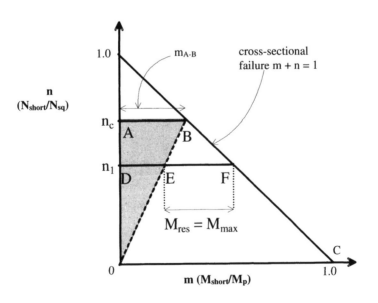

Figure 8.6 Graphical interpretation of moment capacity at a given load for a steel column

$$N_{crit} = \frac{\pi^2 (EI)_e}{L_e^2} \tag{8.17}$$

where the effective flexural rigidity is taken as

$$(EI)_e = E_s I_s + 0.8 E_c I_c \tag{8.18}$$

Presumably the factor 0.8 in Eq. 8.18 is intended to account for concrete cracking and other material nonlinearities in an approximate way.

8.4.1.2 **Example 8.4** *'Buckling load' of a composite column*

The buckling or critical load for the column shown in Figure 7.5(a) and which buckles about the major axis will be determined. As Eq. 8.18 is approximate, we can justifiably ignore the steel in calculating the concrete properties. Hence $I_c = 900^3 \times 500/12 = 30.4 \times 10^9$ mm^4. Similarly for the steel component we may obtain $I_s = (10 \times 700^3/12) + (2 \times 300 \times 15 \times 357.5^2) = 1436 \times 10^6$ mm^4. Assuming that $E_c = 25$ kN/mm^2, $E_s = 200$ kN/mm^2 and that the effective length of the column is $L_e = 25$ m, from Eq. 8.18 $(EI)_e = (200 \times 1436 \times 10^9) + (0.8 \times 25 \times 30.4 \times 10^{12})$ kNmm$^2 = 895 \times 10^{12}$ Nmm^2, and from Eq. 8.17, $N_{crit} = \pi^2 \times 895 \times 10^{12}/25{,}000^2$ N $= 14{,}133$ kN.

8.4.1.3 *Column curve for composite columns*

It was shown in Section 8.2.1.3 that the slenderness ratio $\lambda = L/r$ is proportional to $\sqrt{(N_{sq}/N_E)}$. This may be extended to composite columns of effective length L_e by writing a modified slenderness

$$\lambda^* = \sqrt{\frac{N_{sq}}{N_{crit}}} \tag{8.19}$$

where N_{crit} is given by Eq. 8.17 and N_{sq} is the squash load given by the sum of Eqs. 7.2 and 7.3. The strength N_{col} of a concentrically loaded composite column may be determined from the modified slenderness λ^* and the relevant steel strength curve (curve 'b' in Figure 8.3) of the Eurocode. The slenderness factor χ^* is tabulated in the Eurocode directly as a function of λ^* and the relevant column curve. Hence by analogy with Eq. 8.8 for a steel column,

$$N_{col} = \chi^* N_{sq} \tag{8.20}$$

for a composite column.

8.4.1.4 **Example 8.5** *Strength of a concentrically loaded column*

For the column considered in Example 8.4 that was also analysed in Example 7.1, $\lambda^* = \sqrt{(16{,}275/14{,}133)} = 1.07$. Hence from curve 'b' of the Eurocode, $\chi = 0.58$ and so the column strength of the slender composite member is $0.58 \times 16{,}275 = 9440$ kN which is the maximum concentric load that can be applied to the column.

8.4.2 *Second order effects*

8.4.2.1   General

Second order effects caused by axial compression and bending in composite columns are treated in exactly the same way as was described in Section 8.3.1 for steel columns. Hence, the maximum moment in the slender column is obtained from the maximum end moment M_{ml} and the moment gradient β by amplifying the maximum end moment by $c_m/(1 - N/N_{crit})$, where c_m is given by Eq. 8.16, N_{crit} by Eq. 8.17 and where N is the axial compression.

Example 8.6 *Moment amplification factor for a slender column*

Suppose now that the column in Example 8.4 is subjected to a moment M_{ml} at one end and $M_{ml}/2$ at the other end that bend the column in single curvature. Hence $\beta = -0.5$ and from Eq. 8.16 $c_m = 0.66 + 0.44 \times 0.5 = 0.88$. The maximum moment within the column M_{max} is therefore obtained by amplifying the end moments by the right hand side of Eq. 8.14. If the applied load N = 6000 kN, then this amplification is $0.88/(1 - 6000/ 14,133) = 1.53$, that is $M_{max} = 1.53 \, M_{ml}$.

8.4.3 *Moment capacity for a given load*

8.4.3.1   General

When a composite column is subjected to a given axial compression N, the Eurocode approach allows us to calculate the bending capacity in an analogous manner to that described for steel beam-columns. Instead of the straight line from n = 1 to m = 1 that describes the cross-section strength in Figure 8.6, the cross-section strength is determined using the methods described in Chapter 7 by either first principles, or the rigid plastic method applied in the Eurocode that was described in Section 7.4. We can again illustrate this method graphically in Figure 8.7, where m is the moment M_{short} non-dimensionalized with respect to the pure bending capacity of the composite column M_p, and n is the compressive force N_{short} non-dimensionalized with respect to the squash load N_{sq}.

Under the action of an axial compression, we determine first the column strength described in Section 8.4.1.3, which non-dimensionally is the slenderness parameter χ in the Eurocode[2] and which is shown in Figure 8.7. Following the same technique for steel beam-columns, the line AB is drawn to intersect the strength envelope (m_{short}, n_{short}) at B, and the line OB constructed. The axial force N is non-dimensionalized to produce $n_d = N/N_{sq}$, which clearly must be less than χ. This axial force is plotted on the vertical axis, and the line CF drawn to intersect the strength interaction envelope at F and the line BO at E. The dimensionless moment EF is thus the bending capacity that is allowed at the level of axial compression N.

This approach is a little conservative, and larger capacities than EF are achievable under higher values of moment gradient. This can also be shown to be true for steel beam-columns. The Eurocode allows for this by extending the line OB up the vertical axis to a point G whose dimensionless capacity is χ_n, in which

$$\chi_n = \frac{(1-\beta)}{4}\chi \ \leq \ n_d \tag{8.21}$$

Using the construction in Figure 8.7,

$$\mu = \mu_d - \mu_k \left(\frac{n_d - \chi_n}{\chi - \chi_n} \right) \tag{8.22}$$

and if the section failure envelope is drawn according to Section 7.4 assuming rigid plastic assumptions then

$$M_{res} = 0.9 \mu M_p \tag{8.23}$$

8.4.3.2   **Example 8.7** *Bending capacity of a composite column*

The failure envelope for the column considered in Example 8.6 was determined in Examples 7.2 to 7.4 and is shown in Figure 7.9. At a value of $\chi = 0.58$, the dimensionless point of intersection with the strength envelope developed in Chapter 7 is $\mu_k = 1.05$. It is worth noting that for a composite column the dimensionless moment m can be greater than unity owing to the concavity of the cross-section failure envelope. The point F in Figure 8.7 corresponding to the load $n_d = 6000/16,275 = 0.37$ is $\mu_d = 1.41$ where $N_{sq} = 16275$ kN from Example 7.4. In addition, from Eq. 8.21, $\chi_n = 0.58 \times (1 + 0.5)/4 = 0.22 < 0.37$, so $\chi_n = 0.22$. Hence from Eq. 8.22, $\mu = 1.41 - 1.05 \times (0.37 - 0.22)/(0.58 - 0.22) = 0.97$, and since the cross-sectional strength envelope was generated making rigid plastic assumptions, $M = 0.9 \times 0.97 \times 1666$ kNm $= 1450$ kNm where $M_p = 1666$ kNm from Example 7.2. This moment is the maximum moment in the cross-section, amplified from the end

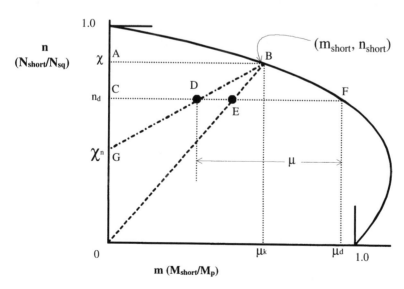

Figure 8.7 Design procedure for compression and bending

moments according to Eq. 8.14. Hence from Example 8.6, the end moments that can be resisted by this column are $1454/1.53 = 950$ kNm and $950/2 = 475$ kNm. These moments correspond to eccentricities of the end load of 6000 kN of $950/6000$ m $= 158$ mm and $158/2 = 79$ mm.

8.5 References

1. Trahair, N.S. and Bradford, M.A. (1998). The Behaviour and Design of Steel Structures to AS4100. 3rd edn, E&FN Spon, London.
2. Eurocode 4 (1994) Design of Composite Steel and Concrete Structures. DDENV 1994-1-1, Draft for development.

9 Composite beams with service ducts

9.1 Introduction

Passing the building services through ducts in the webs of the composite beams in buildings can allow a considerable reduction to the storey height and, hence, reduce the overall height of the building and foundations. The rigid plastic analysis procedures developed in Chapter 4 are used throughout this chapter to assess the effect of inserting a service duct into the web of an existing composite beam and to assess the effect of strengthening the ducted region. It needs to be emphasized that the simplicity of both the ensuing analyses and examples is solely due to the fact that rigid plastic assumptions have been used.

The general analysis procedures[1-4] first developed by Redwood and Darwin are described in Sections 9.2 to 9.6 where a service duct is inserted into an existing beam in a region where flexure predominates. The analysis is then extended in Section 9.7 to determine the enhancement of the strength due to the shear capacity of the slab, and in Section 9.8 to determine the effect of strengthening the ducted region by plating. The procedure is then applied in Section 9.9 to the insertion of a duct in a region of a composite beam where shear predominates. Finally in Section 9.10, methods are proposed for determining whether local embedment failure of the shear connectors will occur due to the insertion of the duct.

9.2 Outline of general analysis procedure

A composite beam with a service duct is shown in Figure 9.1. The duct can be visualized as partitioning the composite beam into the three distinct regions shown, that is the support region A-B, the ducted region B-C and the mid-span region C-D. The section at C will be referred to as the high moment end of the duct and that at D as the low moment end. The insertion of the duct in Figure 9.1 not only weakens the composite beam by reducing the cross-sectional area of steel in the ducted region to $(A_s)_{duct}$ as shown, but can also reduce the longitudinal shear forces in all three regions, that is $(F_{sh})_{A-B}$, $(F_{sh})_{A-C}$ and $(F_{sh})_{A-D}$. The overall flexural capacity of the existing composite beam with a duct and in particular the effect that the duct has on the flexural capacity at the position of maximum moment, is covered in Section 9.3.

The ducted region in Figure 9.1 is shown in Figure 9.2 where it can be seen that the beam now consists of an inverted steel T-section at the bottom of the composite beam and a composite T-section at the top. These T-sections have to resist combinations of the applied shear forces V_a as well as the applied moments M_a. Instead of trying to analyse the ducted region for combinations of these stress resultants that is extremely complex, the following analysis procedure is used:

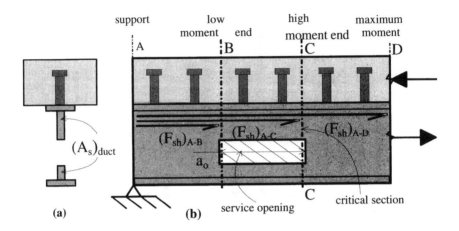

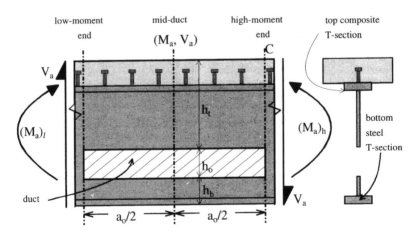

Figure 9.1 Composite beam with service duct

- First of all, the pure flexural capacity of the ducted region is determined in Section 9.4, that is the flexural capacity of the section ignoring the effect of vertical shear.
- The pure shear capacity is then determined in Section 9.5, that is the ability of the T-sections to resist vertical shear in a region of very low applied moment.
- Finally, the interaction between the applied shear V_a and the applied moment M_a is dealt with in Section 9.6, using experimentally derived failure envelopes.

It will be shown in the following analyses that the critical region of the ducted composite beam, in both the ensuing shear analyses and flexural analyses, occurs at the high moment end of the duct at section C in Figures 9.1 and 9.2.

Figure 9.2 Stress resultants acting on ducted region

9.3 Maximum flexural capacity of ducted beam

9.3.1.1 Example 9.1 *Flexural capacity of ducted beam at mid-span*

(a) Effect of transverse ribs

The composite beam in Figure 4.3 in Chapter 4 was previously analysed in Figure 4.4 in Section 4.2.2.2 with full shear connection and with a solid slab. In this analysis, the strength of the steel component was weaker than that of the concrete component and, hence, $P_{sh} = P_s = 2300$ kN at the position of maximum moment. Let us now assume that the solid concrete slab is replaced with a profiled slab with transverse dove-tailed ribs of height 50 mm as shown in Figure 9.3. In the full shear connection analysis in Figure 4.4, the depth of the concrete in compression $n = 30.9$ mm that is less than the depth of 80 mm of the solid concrete above the transverse ribs in Figure 9.3. Therefore, the transverse ribs in the composite beam in Figure 9.3 will not reduce the maximum flexural capacity of the beam in Figure 4.4 and the strength will remain at $M_{fsc} = 702$ kNm.

(b) Insertion of a service duct

Let us now insert a service duct into the beam in Figure 4.3 in the region of high moment near the mid-span of the beam as shown in Figure 9.3. It will be assumed in the following analyses that the stud shear connectors are uniformly distributed along the length of the beam so that the degree of shear connection at any section is directly proportional to the distance of that section from the nearest support. The high moment end of the duct, such as at section C-C in Figure 9.1, is a critical section in the analysis because, at this section, the reduced cross-sectional area of the steel component $(A_s)_{duct}$ is subjected to the largest longitudinal shear force $(F_{sh})_{A-C}$. Hence, let us start the analysis at this section.

At the critical section C-C in Figure 9.3:

- the strength of the concrete component $(P_c)_{C-C} = A_c f_{cy} = 3500 \times 80 \times 21.3$
  $= 5964$ kN

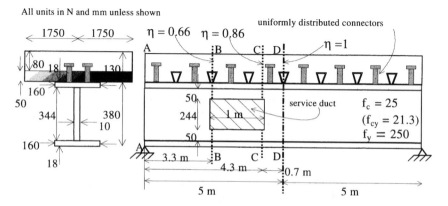

Figure 9.3 Duct inserted into an existing composite beam

- the strength of the steel component $(P_s)_{C-C} = (A_s)_{duct}f_y = 2 \times ((50 \times 10 \times 250)$
  $+ (160 \times 18 \times 250)) = 1690\,kN$
- and as the degree of shear connection $\eta = 4.3/5 = 0.86$,
- the strength of the shear connectors that can impose a thrust on section C-C is
  $(P_{sh})_{C-C} = \eta(P_{sh})_{D-D} = 0.86 \times 2300 = 1978\,kN$ in which from Example 4.1
  $(P_{sh})_{fsc} = 2300\,kN$

The weakest of the three components $(P_c)_{C-C,}$ $(P_s)_{C-C}$ and $(P_{sh})_{C-C}$ that are listed above is the strength of the steel component and, hence, the resultant force in the shear connectors in region A-C in Figure 9.3 is $(F_{sh})_{A-C} = (P_s)_{C-C} = 1690\,kN$. Therefore, the connectors in region A-C are not fully loaded because the weak steel section now controls the design. However, the connectors in region C-D remain fully loaded. As the degree of shear connection within this region C-D is $1 - 0.86 = 0.14$, the strength of the shear connectors within region C-D is $0.14 \times 2300 = 322\,kN$. Therefore, the maximum thrust that the connectors can impose at mid-span is $(F_{sh})_{A-C} + (P_{sh})_{C-D} = 1690 + 322 = 2012\,kN$ that is less than the requirement for full shear connection of $2300\,kN$, so the mid-span now has partial-shear-connection due to the insertion of the duct.

The partial shear connection analysis at mid-span is shown in Figure 9.4. The analysis procedure that is depicted has been fully described in Chapter 4 and in brief consists of:

- The cross-sectional properties of the section shown in (a).
- The axial strengths P of the three components of the composite beam in (b), which are derived from (a) and from the force in the connectors in the shear span under consideration in Figure 9.3 which in this case is the half span.
- The weakest of the three components in (b) controls the force in each component and, hence, the resultant force in each component in (c) is, in this case, the force in the shear connectors of $2012\,kN$.
- An equivalent stress distribution has been used in (d) to simplify the analysis, where the sum of the stresses in the steel component gives the real stress distribution.

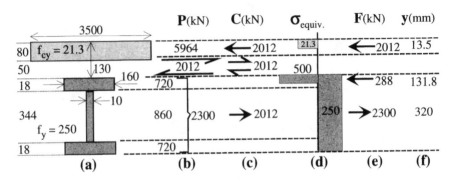

Figure 9.4 Flexural capacity of ducted beam at mid-span

In this case, the neutral axis is in the top steel flange and, hence, the real stress in the steel component above the neutral axis is $500 - 250 = 250$ N/mm² compression and that below the neutral axis is 250 N/mm² tension.

- As the steel component is not fully yielded in tension as shown in (b) and (c) where the force 2012 kN is less than the strength of 2300 kN, part of the steel component must be in compression as shown in (e). The compressive force of 288 kN is such that the sum of forces in the steel component in (e), that is 2300 kN $- 288$ kN, is equal to the resultant force in the steel component in (c) of 2012 kN.
- Knowing the resultant forces in (e), the cross-sectional widths over which they act in (a), and the equivalent stresses in (d), the distance of these resultant forces from the top fibre can be derived and are shown in (f)
- Taking moments about the top fibre using (e) and (f), gives the capacity at mid-span $(M_{psc})_{D-D} = 671$ kNm. This can be compared to the capacity without the duct from Example 4.1 of 702 kNm. Hence, the effect of inserting the duct is to reduce the flexural capacity at mid-span by a small amount of only 4%.

9.4 Pure flexural capacity of ducted region

9.4.1 *Flexural behaviour*

The ducted region in Figure 9.1 is shown subjected to flexure in Figure 9.5. In deriving the flexural capacity within the ducted region, the following standard assumptions are made: the steel and concrete components have the same curvature κ as shown; and a slip-strain $(ds/dx)_{inter}$ exists at the interface between the steel and concrete components. However, in analysing the ducted region for flexure, it will also be assumed that the top steel T-section is in full interaction with the bottom inverted steel T-section, so that the strain profile in the top steel T-section is in line with that of the bottom steel T-section as shown, that is $(ds/dx)_{duct} = 0$. Hence, the rigid plastic analysis procedures developed in Chapter 4 can be applied directly to the composite ducted section.

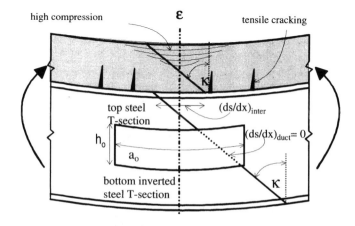

Figure 9.5 Ducted region subjected to flexure

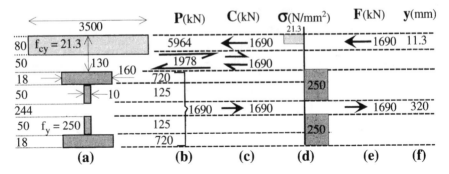

Figure 9.6 Pure flexural capacity at high moment end of duct

9.4.1.1   **Example 9.2** *Pure flexural capacity at high moment end of ducted region*

The analysis for the pure flexural capacity of the ducted region at the high moment end of the duct, at section C-C in Figure 9.3, is shown in Figure 9.6. The strength of the concrete component is listed in (b) as 5964 kN, the strength of the individual rectangular elements of the steel component sums to 1690 kN, and the strength of the shear connectors between the high moment region of the duct and the support is shown as 1978 kN. Hence, the weakest component in (b) is the steel component that, therefore, controls the forces in the components as shown in (c). It can now be seen by comparing (c) with (b) that the connectors in the shear span A-C in Figure 9.3 are now not fully loaded due to the presence of the duct, which will also affect the flexural strength at the low moment end of the duct. As the strength of the steel component controls the force distribution, the steel component is fully yielded in tension as shown in Figure 9.6(d) so that there is no need to use the equivalent stress distribution in this analysis. From (e) and (f), the pure flexural capacity at the high moment end of the duct comes to $(M_{duct})_h = 522$ kNm.

9.4.1.2   **Example 9.3** *Pure flexural capacity at low moment end*

(a) Pure flexural capacity at low moment end

It was shown in Example 9.2 that the shear connectors in shear span A-C in Figure 9.3 are now no longer fully loaded but are resisting a force of $(F_{sh})_{A-C} = 1690$ kN. It will be assumed that this force is uniformly distributed along the shear span A-C so that at the low moment end of the duct at section B-B, the shear connector force $(F_{sh})_{A-B} = (F_{sh})_{A-C}$ $\times L_{A-B}/L_{A-C} = 1690 \times 3.3/4.3 = 1297$ kN. This is shown as the strength of the shear connectors in Figure 9.7(b) in the analysis at the low moment end. From (e) and (f) in Figure 9.7, the pure flexural capacity at the low moment end $(M_{duct})_l = 477$ kNm.

(b) Pure flexural capacity of ducted region

It can be assumed that the pure flexural capacity at mid-span of the duct, which is needed in the analysis in Section 9.6, is the average of that at the high and low

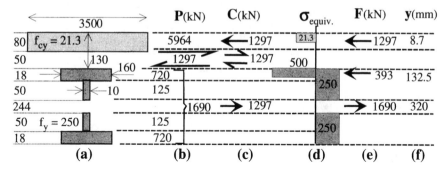

Figure 9.7 Pure flexural capacity at low moment end of duct

moment ends, that is $(M_{pure})_{duct} = [(M_{duct})_h + (M_{duct})_l]/2 = 500$ kNm. Alternatively, the pure flexural capacity can be derived directly following the procedures already outlined for the low moment end.

9.5 Pure shear capacity of ducted region

9.5.1 *Mechanism of shear transfer*

The shear deformation of the ducted region under an applied shear force V_a is shown in Figure 9.8. As the low moment end is subjected to negative or hogging curvature whilst the high moment end is subjected to positive or sagging curvature, it can be seen that a point of contraflexure exists within the ducted region. Hence, the duct spans across a point of zero flexure so the applied moment M_a, that is in the vicinity of the duct in Figure 9.8, will be small and can be ignored in this analysis of the pure shear capacity. It is also worth noting that at the high moment end of the duct at section C-C in Figure 9.8, the concrete is in compression at the top of the slab, however and in contrast, at the low moment end of the duct at section B-B, the concrete is in compression at the soffit of the slab. This transfer of the position of the resultant compressive force in the slab across the ducted region needs to be allowed for in the analysis. Furthermore, it is also worth noting that the shear V_a is transferred across the duct by a mechanism within the bottom steel inverted T-section and also by a mechanism within the top composite T-section.

The mechanism by which the applied shear V_a is transferred across the ducted region in Figure 9.8 is shown in Figure 9.9. In the preceding flexural analysis, the steel T-sections were assumed to act together as shown in Figure 9.5 by $(ds/dx)_{duct} = 0$. In contrast in this pure shear capacity analysis, the T-sections are assumed to act independently of each other. As the applied moment M_a in Figure 9.8 is assumed to be very small, the axial force F in both of the T-sections in Figure 9.9 are, therefore, assumed to be zero as shown.

Let us first consider the bottom steel T-section in Figure 9.9. Equilibrium occurs when

$$M_{bh} + M_{bl} = V_b a_o \tag{9.1}$$

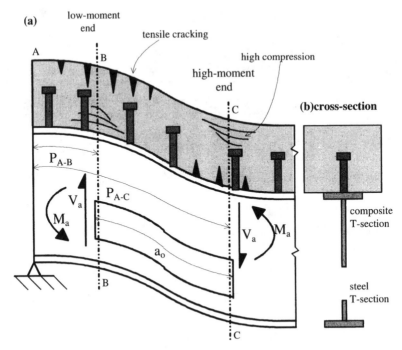

(a)

low-moment end

tensile cracking

high compression

high-moment end

A

B

high-moment end

C

P_{A-B}

V_a

M_a

P_{A-C}

V_a M_a

a_o

B

C

(b)cross-section

composite T-section

steel T-section

Figure 9.8 Deformation of ducted region subjected to large shear forces

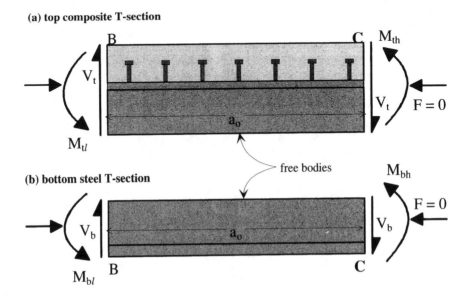

(a) top composite T-section

B   C M_{th}

V_t

M_{tl}

V_t $F = 0$

a_o

free bodies

(b) bottom steel T-section

M_{bh}

V_b $F = 0$

V_b

a_o

M_{bl} B   C

Figure 9.9 Shear transfer mechanism

Hence the shear force V_b induces a moment across the duct of $V_b a_o$ that has to be resisted by local moments at the ends of the T-section of M_{bh} and M_{bl}. It can be seen that the shear force is transferred across the duct by a mechanism of local moments. These moments have been referred to as 'local' as they do not resist the applied moment M_a but are only there to resist the moment induced by the shear forces. As the shear force is increased, the ability of the T-section to resist flexure diminishes as the flexural capacity of the steel material is reduced by the shear stresses as given by Eq. 4.2. The analysis procedure consists of determining the maximum shear force V_b that can be applied, such that the remaining local flexural capacity can just resist the moment induced by this shear force.

For the top composite T-section in Figure 9.9,

$$M_{th} + M_{tl} = V_t a_o \qquad (9.2)$$

The analysis procedure follows the same principles as that of the bottom T-section, but in this case the capacity to resist the shear V depends on the local flexural capacities of the composite T-section that will be shown to have partial shear connection.

The pure shear capacity of the ducted section is then equal to the sum of the individual capacities of the T-sections.

$$V_{pure} = V_t + V_b \qquad (9.3)$$

9.5.2 *Pure shear capacity of steel t-section*

9.5.2.1   **Example 9.4** *Pure shear capacity of bottom steel inverted T-section*

(a) Iterative approach

Von Mises' yield criterion, given in Eq. 4.2, will be used to allow for the interaction between the shear stresses induced by V_b in Figure 9.9 and the flexural stresses induced by M_b. Because of this interaction, the equivalent yield strength of the steel f_{fy} depends on the shear stress τ_w that is unknown at the start of the analysis, so that an iterative approach is required to find a solution. Hence, the first step in the analysis is to guess a reasonable value for the shear force V_b.

(b) Material shear capacity

It will be assumed in these analyses that only the web of the steel section resists shear. Some approaches assume that the extension of the web into the flange can also resist the shear and this can be easily allowed for as shown elsewhere[4]. It will also be assumed in these analyses that the shear stress τ_w is uniformly distributed over the web area A_{web}, that is $\tau_w = V_b/A_{web}$.

The maximum shear stress that the steel material can resist can be derived from Eq. 4.2 as $f_y/\sqrt{3}$ by inserting $f_{fy} = 0$. Hence, the material shear strength of the web is given by

$$\left(V_b\right)_{mat} = \frac{A_{web} f_y}{\sqrt{3}} \qquad (9.4)$$

which is an upper bound to the shear force that can be applied to the T-section. For the bottom inverted steel T-section in Figure 9.3, $(V_b)_{mat} = 50 \times 10 \times 250/\sqrt{3}$ = 72 kN. Hence, let us start our iterative analysis by applying a shear force of 25 kN.

(c) $V_b = 25$ kN

Consider the bottom T-section at the high moment end of the duct in Figure 9.3. For V_b = 25 kN, $\tau_w = 25,000/500 = 50$ N/mm^2 and, hence, from Eq. 4.2 the flexural strength of the web is $f_{fy} = \sqrt{(250^2 - 3 \times 50^2)} = 235$ N/mm^2. The axial strengths of the rectangular elements of the T-section can now be determined as shown in Figure 9.10(b). It can be seen in (c) and (d) that the flexural strength of the steel is now either $f_{fy} = 235$ N/mm^2 or $f_y = 250$ N/ mm^2. The resultant forces are shown in (e) where $602 = P_{flange} - P_{web} = 720 - 118$. As we are dealing with the high moment end of the duct, the forces shown in (e) are those that are external to the ducted region that is shown hatched. From (e) and (f), the external moment to the duct is $M_{bh} = +7.1$ kNm, where the positive sign signifies an anti-clockwise moment.

The analysis depicted in Figure 9.10 can also be applied to the low moment end of the duct. This gives $M_{bl} = +7.1$ kNm, which is exactly the same moment capacity as at the high moment end of the duct because the T-section is steel throughout. Considering equilibrium of the T-section in Figure 9.9(b) gives $M_{bl} + M_{bh} = 14.2$ kNm and $V_b a_o$ $= -25 \times 1 = -25$ kNm. Hence, the moment induced by the shear force $V_b a_o$ exceeds the remaining flexural capacity $M_{bl} + M_{bh}$, so that the shear load has to be reduced.

(d) Graphical representation of results

Equilibrium of the bottom T-section is shown graphically in Figure 9.11. The results of analyses that fall below the 'equilibrium line' are safe as the remaining local flexural capacity $M_{bl} + M_{bh}$ exceeds the moment induced by shear $V_b a_o$. In the preceding analysis, the available local moment capacity is shown as $M_{avl} = 14.2$ kNm and that required to resist the shear is shown as $M_{rqd} = 25$ kNm. Hence in the next analysis, the applied shear load is reduced to 15 kN.

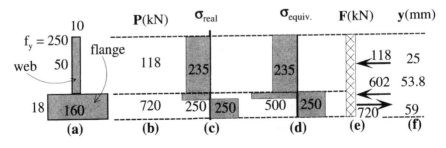

Figure 9.10 Local flexural capacity at high moment end at 25 kN

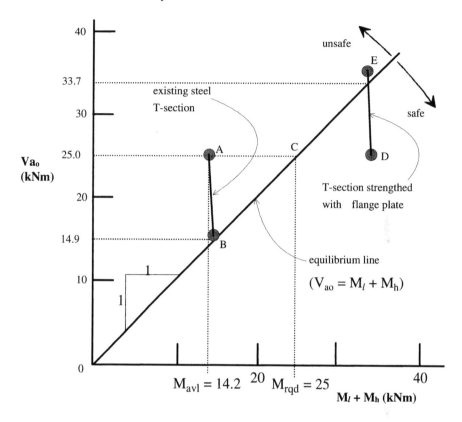

Figure 9.11 Shear capacity of bottom steel T-section

(e) $V_b = 15$ kN

The analysis at a shear load of 15 kN is shown in Figure 9.12. From this analysis $M_{bl} + M_{bh} = 14.7$ kNm and $V_b a_o = 15 \times 1 = 15$ kNm that is shown as point B in Figure 9.11, which is very close to the 'equilibrium line'. Using a linear extrapolation of points A and B to intersect the equilibrium line shows that equilibrium occurs at $V_b a_o = 14.9$ kNm, that is $V_b = 14.9$ kN as $a_o = 1$ m.

9.5.3 *Pure shear capacity of composite t-section*
9.5.3.1 **Example 9.5** *Pure shear capacity of top composite T-section*
(a) Local flexural capacity at high moment end ($V = 50$ kN)

In the beam in Figure 9.3, the duct has been placed at the mid-depth of the web, so that the steel web above and below the duct are the same size. Hence, the material shear capacity of the web of the top T-section is the same as that of the bottom steel section, that is $(V_{mat})_t = 72$ kN. As the top composite T-section is substantially larger that the bottom steel T-section, let us start with a shear load of $V_t = 50$ kN. The

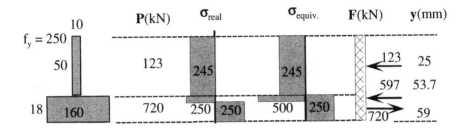

Figure 9.12 Local flexural capacity at high moment end at 15 kN

analysis is shown in Figure 9.13, from which the flexural capacity is $M_{th} = +111.3$ kNm. It is worth noting that at this high moment end, the top portion of the concrete slab is in compression as shown in Figure 9.8 and, hence, standard flexural strength analysis procedures can be followed.

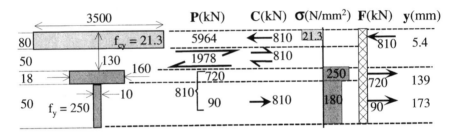

Figure 9.13 Top composite T-section at high moment end at 50 kN

(b) Local flexural capacity at low moment end (V = 50 kN)

It can be seen in the analysis at the high moment end in Figure 9.13 that the shear connectors are not fully loaded but resist a shear load of 810 kN. By assuming that this shear load is uniformly distributed over the shear span A-C in Figure 9.3, the shear connector force in the shear span A-B is $(F_{sh})_{A-B} = 810 \times 3.3/4.3 = 622$ kN, which is shown in the analysis in Figure 9.14 as the strength of the shear connector component in the profile entitled P. This component force controls the resultant forces as shown in profile C.

Care should now be taken at this stage of the analysis. It can be seen in Figure 9.8 that at the low moment end of the duct, the concrete compression zone lies near the bottom of the slab. This can only be achieved with the strain distribution shown in Figure 9.14(d) that requires a very large slip strain and, hence, is peculiar to composite sections with mechanical shear connectors. The real stress distribution associated with this unusual strain distribution is shown in (e). It can be seen that the concrete component is now in compression at the bottom of the solid part of the slab, that is above the transverse ribs in the profiled sheeting. Furthermore, the steel component is now in tension at its top and in compression in its lower parts. The equivalent

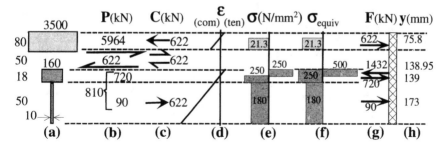

Figure 9.14 Top T-section at low moment end at 50 kN

stress system is shown in (f) and the forces associated with this equivalent stress distribution in (g) where the hatched region signifies the ducted region.

The tensile force of 1432 kN in Figure 9.14(g) is derived from the fact that the resultant force in the steel component in (g) is 622 kN (= 1432 – 720 – 90) as shown in (c). Alternatively it can be derived from the forces in (g) summing to zero, that is 622 + 720 + 90 = 1432 kN. From (f), the equivalent tensile stress in the steel flange is $2f_y = 500$ N/mm² and, hence, the maximum tensile force in the steel flange is $2 \times 720 = 1440$ kN. As the tensile force required of 1432 kN is less than the tensile strength of 1440 kN, the neutral axis must lie in the steel flange as shown in (f). The forces in (g) are the forces external to the low moment end of the duct in Figure 9.8 and, hence, have been shown to the left of the duct region shown hatched. The direction of these forces is extremely important and should be shown as the forces external to the duct. The position of these resultant forces is shown in (h). Taking moments from (g) and (h), with anti-clockwise moments being positive, gives $M_{tl} = -36.2$ kNm, that is the moment at the low moment end is now acting in a clockwise direction. It is worth noting that unlike the local moment at the high moment end M_{th} that remains constant in direction, the direction of the moment at the low moment end can change, as it depends on the position of the duct along the length of the composite beam.

(c) Shear capacity ($V_t = 50$ kN)
From Eq. 9.2, $M_{th} + M_{tl} = 111.3 - 36.3 = 75$ kNm, however, $V_t a_o = 50 \times 1 = 50$ kNm. Hence, the available flexural capacity of 75 kNm is more than sufficient to resist the moment induced by the shear load of 50 kNm. The result is shown as point A in Figure 9.15. In this graphical presentation of the results, the shear load is plotted along the ordinate and the local flexural capacity as the abscissa, so that the slope of the line that represents equilibrium is a function of the length of the duct a_o as shown.

(d) Shear capacity ($V_t = V_{mat} = 72$ kN)
As point A in Figure 9.15 lies below the equilibrium line for the beam with a duct length of $a_o = 1$ m, the shear force can be increased. Let us try the maximum shear load that the steel web can resist, that is $V = V_{mat} = 72$ kN, and hence $f_{fy} = 0$ for the web. The

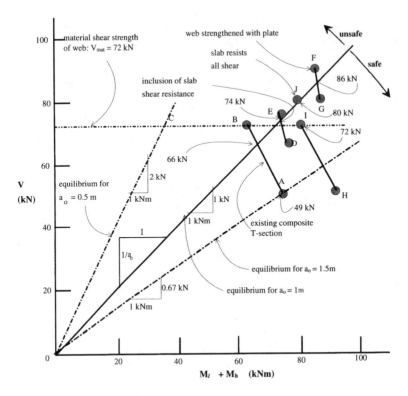

Figure 9.15 Shear capacity of top composite T-section

analysis for the high moment end is shown in Figure 9.16, where it can be seen that the strength of the web shown in the strength profile P is shown as zero. The local moment capacity comes to $M_{th} = +96.6$ kNm

It can be seen in Figure 9.16 that at the high moment end of the duct the shear connectors are not fully loaded but resist a force of 720 kN. Therefore at the low moment end of the duct, the force exerted by the shear connectors $(F_{sh})_{A-B}$ = $720 \times 3.3/4.3 = 553$ kN, as shown in the analysis of the low moment end in Figure 9.17 from which $M_u = -33.4$ kNm.

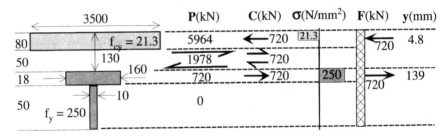

Figure 9.16 Top composite T-section at high moment end at 72 kN

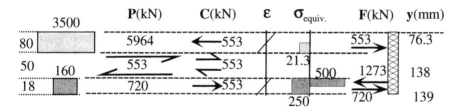

Figure 9.17 Top composite T-section at low moment end at 72 kN

From the preceding calculations $M_{th} + M_{tl} = 96.6 - 33.4 = 63.2$ kNm when $V_t a_o = 72$ kNm. This coordinate is shown as point B in Figure 9.15. The linear interpolation between points A and B intersects the equilibrium line at $V_t = 66$ kN that is, therefore, the shear capacity of the top composite T-section. It is worth noting that the shear capacity of the bottom steel T-section, which had an identical steel component as the top T-section and hence an identical V_{mat}, was previously shown to be 15 kN. The increase in the shear capacity from 15 kN to 66 kN is due to the composite action of the top steel T-section with the slab and has nothing to do with the shear capacity of the slab which was not considered in this analysis; the contribution of the shear capacity of the slab is considered in Section 9.7.

(e) Pure shear capacity of ducted region

From Example 9.4, the pure shear capacity of the bottom T-section was derived as $V_b = 15$ kN, whereas, from Example 9.5 the pure shear capacity of the top T-section is $V_t = 66$ kN. Hence, the pure shear capacity of the ducted region $(V_{pure})_{duct} = 15 + 66 = 81$ kN.

(f) Variation in the duct size a_o

The variation A-B in Figure 9.15 was derived for a duct size of $a_o = 1$ m. Increasing the duct size to $a_o = 1.5$ m will lower the equilibrium line as shown. Although the analyses that were used to determine the results A and B are not directly applicable to a duct size of 1.5 m, as they were derived for a duct size of 1 m, the linear extrapolation of these points to intercept the new equilibrium line of $a_o = 1.5$ m will give the engineer a very good indication of the effect of increasing the duct size by 0.5 m. It can be seen that increasing the duct size from 1 m to 1.5 m will reduce the pure shear capacity from 66 kN to 49 kN.

In contrast, reducing the duct size to 0.5 m in Figure 9.15 will raise the equilibrium line as shown and, thereby, allow increased shear loads. As point B was determined at $V_{mat} = 72$ kN, that is at the largest shear force that can be applied to the web of the composite T-section, the applied shear force cannot be increased to find a point that intercepts the new equilibrium line. In fact it is not necessary, as point B lies on the safe side of the equilibrium line at $a_o = 0.5$ m, that is the local restraining moment capacity of 63 kNm (point B) is more than adequate to resist the moment induced by shear of $V_t a_o = 72 \times 0.5 = 36$ kNm (Point C).

9.6 Interaction between shear and flexure

9.6.1 *Failure envelope*

The analyses in Section 9.4 dealt with the derivation of the pure flexural capacity of the ducted region M_{pure} and that in Section 9.5 derived the pure shear capacity V_{pure}. These results can be visualized are just two points on the extremities of a failure envelope of the combination of stress resultants acting at a section that cause failure, as shown at points A and B in Figure 9.18.

In a statically determinate beam, as the load is gradually increased the combination of the applied stress resultants at the ducted region M_a and V_a follows the linear load path O-C in Figure 9.18. The intercept of this load path O-C with the failure envelope A-C-B at point C is the combination of the stress resultants M_{int} and V_{int} that causes failure. Darwin[2] derived the following failure envelope experimentally.

$$\left(\frac{M_{int}}{M_{pure}} \right)^3 + \left(\frac{V_{int}}{V_{pure}} \right)^3 = 1 \tag{9.5}$$

which for a statically determinate beam can be written in the following form

$$M_{int} = M_{pure} \left[\frac{\left(\dfrac{M_{pure}}{V_{pure}} \right)^3}{\left(\dfrac{M_a}{V_a} \right)^3} + 1 \right]^{-\frac{1}{3}} \tag{9.6}$$

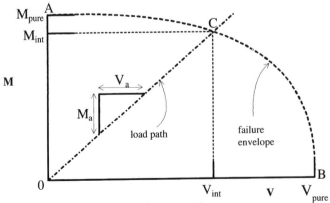

Figure 9.18 Failure envelope of stress resultants

where M_a and V_a are the applied moment and shear force at the ducted region and M_{int} is the flexural capacity of the ducted region that allows for the reduction in strength due to V_{int}.

9.6.1.1 **Example 9.6** *Flexural capacity of ducted region*

The mid-span of the duct in Figure 9.3 occurs at a distance of $L_{duct} = 3.8$ m from the nearest support. In Examples 9.2 and 9.3, the pure flexural capacity at $L_{duct} = 3.8$ m was derived as $(M_{pure})_{duct} = 500$ kNm and in Examples 9.4 and 9.5, the pure shear capacity was derived as $(V_{pure})_{duct} = 81$ kN. Let us assume that the simply supported beam in Figure 9.3 is supporting a uniformly distributed load w_a kN/m. From simple statics, the ratio of the applied stress resultants at $L_{duct} = 3.8$ m is given by $(M_a)_{duct}/(V_a)_{duct} = 11.7w_a/1.2w_a = 9.8$ m. Applying these results to Eq. 9.6 gives $M_{int} = 0.93M_{pure} = 0.93 \times 500 = 465$ kNm. It can be seen that the shear force at the ducted region has reduce the flexural capacity at the ducted region from 500 kN to 465 kNm, that is by 7%.

9.7 Enhanced shear strength due to the shear resistance of the slab

9.7.1 *Contribution of slab*

In the derivation of the pure shear capacity of the top composite T-section in Example 9.5, it was assumed that the web of the steel component resisted all the vertical shear force, which is definitely a safe assumption but probably also a reasonable assumption when the slab does not have longitudinal reinforcing bars. Let us now assume that the slab has at least some nominal longitudinal reinforcing bars, so that the slab can now also resist vertical shear forces. The simplest solution to determining the enhanced shear capacity of the ducted region due to the shear resistance of the slab V_c is to assume that the slab reduces the shear in the steel web of the top composite T-beam to V_s as shown in Figure 9.19.

It can be seen in Figure 9.19 that $V_t = V_c + V_s$. The composite T-section has still to resist the moment induced by shear $V_t a_o$ through the local moments $M_{th} + M_{tl}$. The effect of resisting at least part of the vertical shear by the slab, is to reduce the shear stress in the web and, thereby, increase its effective flexural strength f_{fy} that will increase the local moment capacity and, thereby, increase the overall shear capacity.

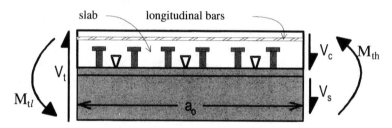

Figure 9.19 Shear resisted by slab in top composite T-section

9.7.1.1 **Example 9.7** *Slab resists 26 kN of the vertical shear force*

Let us now determine the pure shear capacity of the beam in Figure 9.3 by allowing for the enhancement in strength due to the shear resistance of the slab. The first problem is to estimate the shear resistance of the slab.

(a) Shear resistance of slab over ducted region

It would appear to be unreasonable to assume that the full width of slab of 3.5 m in Figure 9.3 could transfer the shear across the ducted region of only 1 m length. Instead, let us just consider the top composite T-section in Figure 9.19 as acting as a beam of span $L = a_o = 1$ m. For guidance, we can use methods for estimating the effective width of beams w_{eff}, such as those described in Section 2.2.2, for estimating the effective width of this composite T-beam of span a_o. For example, codes often use $w_{eff} \approx 0.25L$, as in Eq. 2.3, which in this example gives $(w_{eff})_{duct} = 0.25 \times 1000 = 250$ mm.

The shear resistance of the width of slab of $(w_{eff})_{duct} = 250$ mm can be derived from the shear resistance of beams or slabs without stirrups as given in national standards. Alternatively and as a guideline, the shear strength of an initially cracked section that is given by Eq. 6.1 could be used, as in the following example:

- The weakest section of the slab in Figure 9.3 occurs above the profiled sheeting where the depth of the solid slab is $d_{solid} = 80$ mm deep.
- Hence, the area of slab that resists the vertical shear $A_{slab} = (w_{eff})_{duc} \times d_{solid} = 250 \times 80 = 20,000$ mm².
- The minimum shear resistance from Eq. 6.1 is $0.66f_{ct}$. Hence for $f_{ct} = 2$ N/mm², the minimum shear resistance of the slab $V_c = 0.66 \times 2 \times 20,000 = 26.4$ kN.
- It should be noted that to achieve the minimum shear resistance of $0.66f_{ct}$ in Eq. 6.1, a minimum strength of bars[4] crossing the shear plane is required of $pf_{yr} = 0.66f_{ct}$, where p is the cross-sectional area of the reinforcing bars as a proportion of the area of the shear plane. For $f_{yr} = 400$ N/mm² and $f_{ct} = 2$ N/mm², the minimum requirement gives $p = 0.33\%$.
- Hence $V_c \geq 26.4$ kN when there is at least 0.33% of longitudinal reinforcing bars in the solid portion of the slab.

(b) Pure shear capacity of top composite T-section

Let us assume that the slab over the ducted region in Figure 19.9 can resist $V_c = 26$ kN, and let us start our iterative analysis with $V_s = 50$ kN so that $V_t = 50 + 26 = 76$ kN and, hence, $V_t a_o = 76 \times 1 = 76$ kNm. As $V_s = 50$ kN, the analyses in Figures 9.13 and 9.14 apply directly and, hence, $M_{th} + M_{tl} = 111.3 - 36.2 = 75$ kNm. The results are plotted as point E in Figure 9.15 and as this point lies above the equilibrium line, we will reduce the shear force at our second attempt to $V_s = 40$ kNm. The second analysis at the high moment end is depicted in Figure 9.20 and that at the low moment end in Figure 9.21.

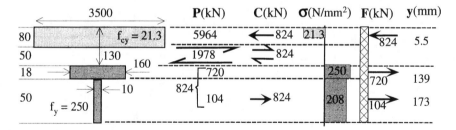

Figure 9.20 High moment end at V_s = 40 kN

From Figure 9.20, $M_{th} = +113.5$ kNm and from Figure 9.21 (where it has been assumed that the neutral axis is at the web/flange interface because the required force of 1456 kN is very close to the 'flange strength' of 1440 kN), $M_u = -36.6$ kNm and, hence, $M_{th} + M_u$ = 76.9 kNm. As $V_s = 40$ kN and $V_c = 26$ kN, $V_t = 66$ kN and $V_t a_o = 66 \times 1 = 66$ kNm. The results are plotted as point D in Figure 9.15. The linear interpolation of points E and D intersects the equilibrium line at $V_{pure} = 74$ kN. It can be seen that the inclusion of the shear resistance of the slab of 26 kN has increased the pure shear capacity by 8 kN, from 66 kN to 74 kN. Hence, the analysis is not sensitive to the shear capacity of the slab. This is because no matter what the shear capacity of the slab is, the moment induced by the shear force $V_t a_o$ has still to be resisted by the local moments $M_{th} + M_u$ at the ends of the duct. This phenomenon is further illustrated in the next example where it is assumed that the slab resists all of the shear force.

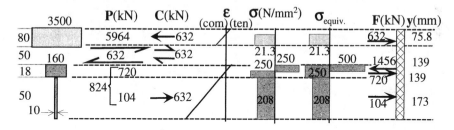

Figure 9.21 Low moment end at V_s = 40 kN

9.7.1.2 **Example 9.8** *Slab resists all the vertical shear force*
If the slab over the ducted region in Figure 9.3 resists all of the vertical shear, then shear stresses will not be present in the web of the top T-section, which will now have its full yield strength of $f_{fy} = f_y = 250$ N/mm². The analyses for the local moment capacities are shown in Figures 9.22 and 9.23 from which $M_{th} = +116.9$ kNm and $M_u = -37.3$ kNm that gives a local moment capacity of $M_{th} + M_u = 80$ kNm. Equating this local moment capacity to the moment induced by the shear gives $M_{th} + M_u = 80 = V_t a_o$. As $a_o = 1$ m, this gives the upper bound to the pure shear capacity of $(V_t)_{upper} = 80$ kN.

The upper bound to the shear resistance, $(V_t)_{upper} = 80$ kN, is shown as point J in Figure 9.15. Hence, the maximum increase in the pure shear capacity is $80 - 66 = 14$ kN. This maximum increase cannot be exceeded no matter how strong

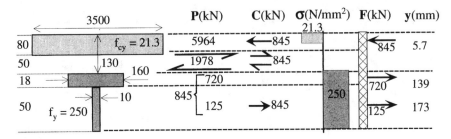

Figure 9.22 High moment end when slab resists all the shear

the shear resistance of the slab. In this example, the maximum shear strength required of the slab is 80 kN as any further increase would not increase the pure shear capacity of the ducted region. It is worth emphasizing further, that even if the slab could resist all the shear load of 80 kN, the maximum increase in the pure shear capacity is only 14 kN that is only 21% of the shear capacity with the web taking all of the shear. Hence, the analysis is not sensitive to the shear capacity of the slab so that any reasonable estimate should suffice.

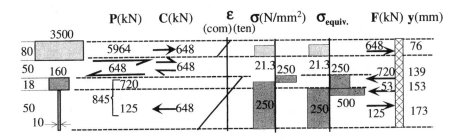

Figure 9.23 Low moment end when slab resists all the shear

9.8 Strengthening ducted regions by plating

9.8.1 *Plating*

If the previous calculations had shown that inserting a duct would not allow the ducted beam to withstand the required design loads, then the ducted region could be strengthened by welding or bolting steel plates to the steel beam, as illustrated in the following examples.

9.8.1.1 **Example 9.9** *Flange plate attached to bottom steel T-section*

A plate can be welded or bolted to the bottom flange to increase the pure shear capacity of the bottom steel T-section, such as shown in Figure 9.24. In this example, the area of the attached flange plate is equal to the area of the bottom flange. The analyses at two values of shear load, $V_b = 25$ kN and at $V_b = 35$ kN, are shown and they are applicable to both the high and low moment ends. At $V_b = 25$ kN, $M_{bh} + M_{bl}$

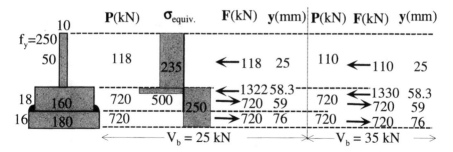

Figure 9.24 Bottom T-section strengthened with an additional flange plate

$= 34.4$ kNm and at $V_b = 35$ kN, $M_{bh} + M_{bl} = 33.8$ kNm. It is worth noting that there is only a very small reduction in the local moment capacity from 34.4 kNm to 33.8 kNm (that is 2%) when the shear load is increased from 25 kN to 35 kN (that is 40%). This emphasizes the insensitivity of the local flexural capacity to the shear force assumed to be resisted by the steel webs.

The results of the analyses in Figure 9.24 are shown as points D and E in Figure 9.11. It can be seen that the addition of the flange plate has substantially increased the pure shear capacity of the bottom steel T-section from 14.9 kN (at $V_b a_o$ = 14.9 kNm) to 33.7 kN (at $V_b a_o = 33.7$ kNm).

9.8.1.2 **Example 9.10** *Flexural capacity at mid-span*

It was shown in Example 9.1 that the insertion of the duct into the beam in Figure 9.3 would reduce the flexural capacity at mid-span by 4%. This reduction can be overcome, if required, by attaching a plate to the flange. In this example, the cross-sectional area of the plate that is to be attached to the flange needs to be greater than or equal to the cross-sectional area of the plate that was removed from the web to form the duct, such as that shown in Figure 9.24. Furthermore, the plate should be placed over the full length of the ducted region. The addition of this plate will allow all the connectors in the composite beam in Figure 9.3 to be fully loaded and, hence, it will maintain the flexural capacity at mid-span.

9.8.1.3 **Example 9.11** *Web plates attached to top composite T-section*

The addition of web plates has been used in Figure 9.25 to increase the pure shear capacity of the top composite T-section. The additional web plate is of equal area to the original web plate. The analysis for the high moment end is given in Figure 9.25 and that at the low moment end in Figure 9.26.

At $V_t = 90$ kN ($V_t a_o = 90$ kNm), $M_{th} + M_{tl} = 128.2 - 41.7 = 86.5$ kNm, and at $V_t = 80$ kN ($V_t a_o = 80$ kNm), $M_{th} + M_{tl} = 130.3 - 42.6 = 87.7$ kNm. These results are plotted as points F and G in Figure 9.15. It can be seen that the addition of the side webs has substantially increased the pure shear capacity from 66 kN to 86 kN.

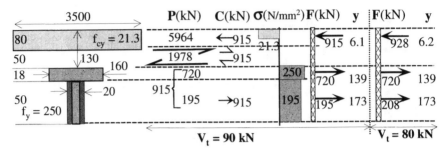

Figure 9.25 Additional web plates, high moment end

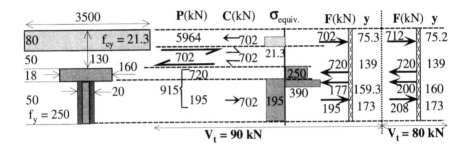

Figure 9.26 Additional web plates, low moment end

9.9 Service duct near supports

9.9.1 *General*

The duct in Figure 9.3 was placed in a high moment region, let us now look at the effect of placing the duct in a low moment region as shown in Figure 9.27. In this region, shear is more likely to predominate than flexure.

9.9.1.1 **Example 9.12** *Flexural capacities*

The composite beam in Figure 9.27 was originally designed with full shear connection in which $P_{sh} = 2300$ kN. The degree of shear connection at the high moment end of the duct is $\eta = 1.7/5 = 0.34$ and, hence, the strength of the shear connection at the high moment end is $(P_{sh})_{A-C} = 0.34 \times 2300 = 782$ kN. Furthermore, it has been shown in previous examples such as in Figure 9.6, the strength of the steel component $(P_s)_{duct}$ = 1690 kN. As the strength of the steel component in the ducted region $(P_s)_{duct}$ is greater than $(P_{sh})_{A-C}$, the connectors are fully loaded in shear span A-B and, hence, they are fully loaded throughout the length of the beam. Therefore, the flexural capacities in shear span C-D are not affected by the insertion of the duct and remain unchanged. As the connectors are fully loaded throughout the length of the beam, standard rigid plastic analysis procedures described in Chapter 4 can be used to predict the pure flexural capacities throughout the shear span A-D.

All units in N and mm unless shown

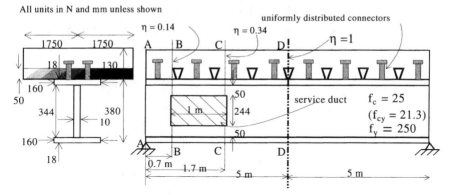

Figure 9.27 Service duct in low moment region

9.9.1.2 **Example 9.13** *Pure shear capacity*

(a) Bottom T-section

The pure shear capacity of the bottom T-section is not affected by the position of the duct and, hence, it is given by the analyses depicted in Figures 9.10 and 9.12 with the results plotted in Figure 9.11.

(b) Top T-section

The analyses for the pure shear capacities of the top composite T-section at the high moment end and at a shear load of 50 kN is shown in Figure 9.28, from which $M_{th} = +107.9$ kNm. It can also be seen in Figure 9.28 that the connectors are fully loaded, so they are also assumed to be fully loaded in the analysis of the low moment end in Figure 9.29, where $(P_{sh})_{A-B} = (0.7/5)(P_{sh})_{A-D} = 0.14(P_{sh})_{A-D} = 0.14 \times 2300 = 322$ kN. From Figure 9.29, $M_{ul} = -15.2$ kNm. Hence, $M_{th} + M_{ul} = 107.9 - 11.6 = 96.3$ kNm when $V_t a_o = 50$ kNm, and this coordinate is shown as point H in Figure 9.15. As this point lies on the safe side of the equilibrium line, the shear force has been increased to V_{mat} in the next attempt at equilibrium.

In this second attempt, let us apply the maximum shear force of $V_{mat} = 72$ kN. The analyses are shown in Figures 9.30 and 9.31 where it is worth noting that the

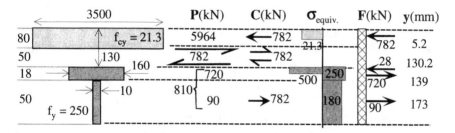

Figure 9.28 Pure shear capacity at high moment end at V = 50 kN

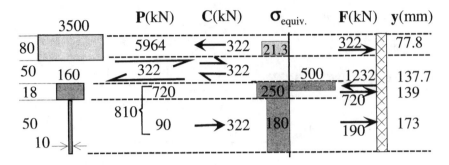

Figure 9.29 Pure shear capacity at low moment end at V = 50 kN

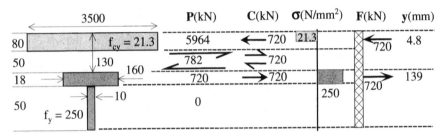

Figure 9.30 Pure shear capacity at high moment end at V_{mat}

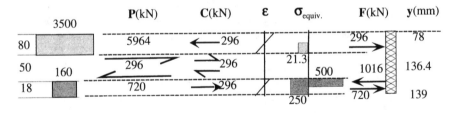

Figure 9.31 Pure shear capacity at low moment end at V_{mat}

connectors are now not fully loaded at the high moment end (720 kN in Figure 9.30) and, hence, the force in the connectors at the low moment end (Figure 9.31) is $(0.7/1.7) \times 720 = 296$ kN. From Figures 9.30 and 9.31, $M_{th} + M_{tl} = 96.6 - 15.4 = 81$ kNm when $V_a = 72$ kN, and this result is shown as point I in Figure 9.15. It can be seen that even when the maximum shear is being applied, the local moment capacity is sufficient, so that the pure shear capacity is $V_{mat} = 72$ kN.

(c) Reversal in moment direction at low moment end
In the analyses of the ducted beam in Figure 9.3, shown in Figures 9.14 and 9.17, the local moment capacity at the low moment end acted in a clockwise direction and had

a magnitude of $M_{tl} \approx -34$ kNm. When the duct was moved towards the supports as in Figure 9.27, the local moment capacity, as shown in the analyses in Figure 9.29, still acted in a clockwise direction but reduced to $M_{tl} \approx -13$ kNm. It can be seen that as the low moment end of the duct shifts towards the supports, the magnitude of the moment at the low moment end diminishes.

As the low moment end of the duct approaches the support, the force in the shear connectors over shear span A-B in Figure 9.27, $(F_{sh})_{A-B} \rightarrow 0$. The analysis when $(F_{sh})_{A-B} = 0$, is shown in Figure 9.32. There is now no longer any force in the concrete as the resultant force in each component $C = 0$, so that the strain distribution now induces a positive moment, which is in the reverse direction to those previously calculated. The positive moment is now benefiting the transfer of shear as can be seen in Figure 9.9 and Eq. 9.2, where a positive M_{tl} assists the positive M_{th} to transfer $V_t a_o$.

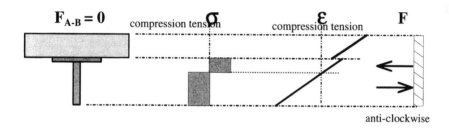

Figure 9.32 Reversal in moment at low moment end of duct

9.10 Embedment failure

9.10.1   *General*

The shear distortion of the ducted region of the composite beam that is shown in Figure 9.8 causes the concrete component to lift away from the steel component at the high moment end of the beam. This separation induces tensile axial forces in the shear connectors that can cause them to pull out of the slab. Another way of visualizing this problem is to consider the longitudinal forces in the slab in the top c   o   m   p   o   s   i   t   e T-section in Figure 9.8 when it is subjected to pure shear, as shown in Figure 9.33.

The longitudinal compressive force H at each end of the ducted region is shown in Figure 9.33. At the high moment end of the duct, H acts within the upper regions of the solid portion of the slab of depth d_{solid} at point A, whereas, at the low moment end of the duct H acts in the lower regions of the solid slab at point B. The transfer of the longitudinal compressive force H from A to B has to be balanced by the normal forces shown as N_{pure}.

At the low moment end of the duct in Figure 9.33, the normal force N_{pure} induces compression across the interface, between the steel and concrete component, which is resisted by the slab bearing against the flange. However, the tension force across the interface at the high moment end can only be resisted by axial tension in the shear

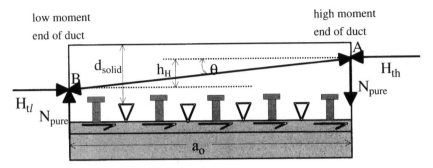

Figure 9.33 Embedment forces in shear connectors

connectors adjacent to the high moment end. The resultant normal force across the interface is zero and, hence, the overall or global longitudinal shear strength of the shear connectors in the composite beam is unaffected. However, the interface tensile force at the high moment end may cause local embedment failure of the shear connectors, which would destroy the composite action of the top T-section and, thereby, destroy its ability to resist shear. In fact, embedment failure would reduce the pure shear capacity of the top composite T-section to that of the top steel T-section.

9.10.1.1 **Example 9.14** *Embedment failure of duct in high moment region*

Let us consider the duct in Figure 9.3 and the longitudinal forces within the duct as shown in Figure 9.33. An estimate of the longitudinal force at point A can be derived from Figures 9.13 and 9.16, from which $H_{th} \approx 745$ kN and acts at $y \approx 5$ mm from the top fibre. Let us now assume that there are no shear connectors within the ducted region in Figure 9.33, so that at the low moment end $H_{tl} = H_{th} = 745$ kN and, therefore, acts at 5 mm from the bottom fibre of the solid concrete slab. Then from simple geometry of the forces within the solid slab of depth 80 mm and length 1000 mm, $\tan\theta = h_H/a_o = (80 - 5 - 5)/1000 = 0.070$. From simple statics at node A, the normal tensile force N_{pure}, which is the normal force when the duct is subjected to the pure shear capacity V_{pure}, is given by $N_{pure} = H_{th}\tan\theta = 52$ kN.

The tensile normal force N_{pure} is the maximum tensile force that occurs when the maximum shear force V_{pure} is acting. In a composite beam, the ducted region is subjected to an applied shear load of V_a. If N_a is the normal tensile force induced by V_a, then as a guide to the magnitude of N_a it is suggested that $N_a = (V_a/V_{pure}) \times N_{pure}$. From Example 9.6, $M_{int} = 465$ kNm which would allow a uniformly distributed load of $w_a = 39.5$ kN/m to be applied to the composite beam. For this applied load, the shear force at the duct $V_a = 47.5$ kN. As $V_{pure} = 81$ kN, then $N_a = (V_a/V_{pure}) \times N_{pure} = (47.5/81) \times 52 = 30$ kN and it is necessary to ensure that the connectors adjacent to the high moment region can resist this embedment force. Shear connectors in solid slabs are usually shaped to have a high resistance to embedment failure and can usually safely resist axial forces of up to 25 % of the shear strength[4]. However, profiled slabs are more prone to embedment failure particularly if the concrete element is a profiled slab with trapezoidal ribs transverse to the beam as shown in Figure 5.3.

9.10.1.2 **Example 9.15** *Embedment failure of duct in high shear region*

Consider now the ducted beam in Figure 9.27 that is subjected to a uniformly distributed load so that the duct is in a low moment and high shear region. The forces in Figure 9.28 can be used as an estimate of the longitudinal compressive force in the slab, from which $N_{pure} = 55$ kN and which is virtually the same as that of the duct in the low moment region in Example 9.14.

If the beam is subjected to the same uniformly distributed load as in Example 9.14, the applied shear force at the duct $V_a = 150$ kN that exceeds the pure shear capacity of $V_{pure} = 72 + 15 = 87$ kN. The ducted beam could be strengthened by plating the flange and the webs as in Examples 9.9 and 9.11 and this would increase the pure shear capacity to $V_{pure} = 86 + 34 = 120$ kN which would still require the applied load to be reduced to $w_a = (120/150) \times 39.5 = 31.6$ kN/m. If the load were reduced, then $V_a = V_{pure}$ so that $N_a = N_{pure} = 55$ kN. It can be seen that shifting the duct towards the supports increases the probability of embedment failure.

9.11 References

1.   Cho, S.H. and Redwood, R.G. (1986). 'The design of composite beams with web openings'. Structural Engineering Series No. 86-2, Department of Civil Engineering and Applied Mechanics, McGill University, Montreal, Canada, June.
2.   Darwin, D. (1988). 'Draft -Design of steel and composite beams with web openings'. American Iron and Steel Institute, U.S.A.
3.   Redwood, R.G. and Poumbouras, G. (1984). 'Analysis of composite beams with web openings'. Journal of Structural Engineering, ASCE, Vol. 110, No. 9, 1949-58.
4.   Oehlers, D. J. and Bradford, M. A. (1995) Composite Steel and Concrete Structural Members: Fundamental Behaviour. Pergamon Press, Oxford.

10 Local splitting

10.1 Introduction

Splitting is probably the most common form of shear connection failure in a composite beam, and often occurs where there is only a small amount of side cover to the shear connectors, such as in composite L-beams and in composite haunched beams. The mechanisms that cause splitting are described first in this chapter. Fundamental analysis techniques are then applied to the derivation of the splitting resistance of a wide variety of the cross-sectional shapes of the concrete component of the composite beam. If it is found that splitting has occurred or is likely to occur, then the post splitting dowel strength of the shear connectors can be estimated from Chapter 11.

10.2 Mechanisms of splitting

Each individual shear connector in a composite beam, such as the stud shear connector in Figure 10.1, imposes a highly concentrated load P_{sh} onto the concrete component that is dispersed longitudinally, vertically and transversely as shown by the arrows marked D. The ability of the slab to resist the longitudinal component of the dispersal, through the formation of the diagonal shear cracks that are shown in (b), has already been dealt with in Chapter 6. However, the transverse dispersal D of the concentrated load in (b) requires the transverse tensile force T to maintain equilibrium. This transverse force can cause a longitudinal split along the line of connectors that can reduce the dowel strength of the shear connectors.

The shear connector in Figure 10.1 can be visualized as a patch load that is being applied to a concrete prism, as shown in Figure 10.2. When the connector is close to the transverse edge of the concrete slab as in Figure 10.1(b), the dispersal of the dowel force induces the transverse stresses in Figure 10.2(a). This is a well known stress distribution[1] that is often used in the analysis of the anchorage zones of post-tensioned members. When the connector is placed well away from the transverse edge, as is the case for most shear connectors, then the transverse stress distribution

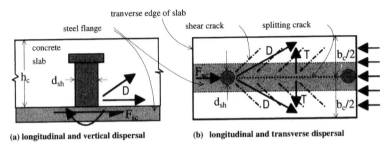

(a) longitudinal and vertical dispersal   (b) longitudinal and transverse dispersal

Figure 10.1 Dispersal of concentrated load

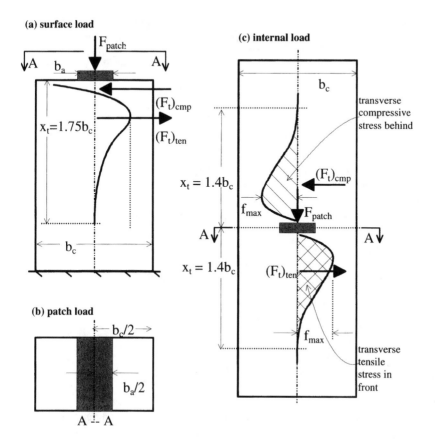

Figure 10.2 Transverse stress distributions

in (c) occurs, that is peculiar to mechanical shear connectors[2] in composite beams. It can be seen in (c) that the transverse tensile stresses in front of the shear connector are balanced by an identical distribution of compressive stresses behind the connector, so that the resultant of the transverse forces $(F_t)_{cmp}$ and $(F_t)_{ten}$ is zero. It can also be seen that the length of the transverse tensile and transverse compressive stress distributions x_t is a function of the width of the prism b_c.

A single longitudinal line of connectors in a composite beam is shown in Figure 10.3. Let us assume that there is a linear variation in the shear flow q as shown, so that the connector force $(F_{sh})_1$ is greater than $(F_{sh})_2$, in which case the lateral stresses will reduce to the left as shown. Bearing in mind that the extent of the transverse stress x_t in Figure 10.2 is a function of the width of the prism b_c, it can be seen in Figure 10.3 that when the longitudinal spacing of the connectors $L_{con} \gg b_c$, then the interaction between the transverse tensile and compressive stresses is minimum. In this case, splitting is a function of the forces in an individual connector that is referred to as *local splitting*.

When in Figure 10.3, $L_{con} \ll b_c$, then there is substantial overlap of the transverse stress distributions of the individual connectors. This interaction between the local stress distributions of the individual connectors can reduce the overall transverse tensile stress in parts of the composite beam but it can also increase the tensile stresses in other parts. This interaction is referred to as *global splitting* as it depends on the distribution of the connectors throughout the beam, and is generally only a problem when the beam is subjected to longitudinally moving loads.

Local splitting of flexible stud shear connectors is dealt with in this chapter. Local splitting of stiff connectors, such as block connectors, and the global splitting of all types of connectors are dealt with elsewhere[2].

10.3 Splitting resistances of slabs with rectangular cross-sections

10.3.1 *Splitting resistance to individual connectors*

In Figure 10.1, a stud shear connector of shank diameter d_{sh} is acting concentrically on the concrete element or concrete prism of a composite beam of width b_c and height h_c. The mean splitting resistance[2] of the concrete prism to the dowel force from an individual stud shear connector is given by the following equation.

$$P_{one} = 3.4 d_{sh}^2 f_{cb} \left(\frac{b_c}{d_{sh}\left(1 - \frac{d_{sh}}{b_c}\right)^2} + \left[\left(1 - \frac{0.9d_{sh}}{(h_c)_{\le 8.1 d_{sh}}}\right)^2 \frac{0.9d_{sh}}{(h_c)_{\le 8.1 d_{sh}}} \right]^{-1} \right) \quad (10.1)$$

where f_{cb} = Brazilian tensile strength of the concrete that can be assumed to be equal to $0.5\sqrt{f_c}$ when f_c is measured in N/mm[2] as suggested in Section 1.3.5 . Furthermore, the characteristic splitting strength, in which 5% of the test results fall below, can be obtained from Eq. 10.1 by substituting the coefficient 3.4 with 2.6.

The first term in the bracket on the right hand side of Eq. 10.1 gives the resistance to splitting of the prism when the concentrated load is dispersed in the longitudinal

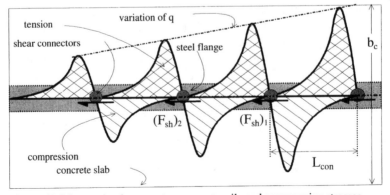

Figure 10.3 Interaction between transverse tensile and compressive stresses

and transverse directions, that is in two-dimensions only as shown in Figure 10.1(b). It can be seen in this term that the splitting resistance increases with the width of the prism b_c. The second term in the bracket gives the increase in the splitting resistance due to the vertical dispersal of the concentrated dowel force into the third dimension as pictured in Figure 10.1(a). It should be noted that in this second term, an upper bound of $8.1d_{sh}$ is placed on the height of the prism h_c to limit the increase in strength due to the vertical dispersal; when $h_c > 8.1d_{sh}$ then it can be assumed that $h_c = 8.1d_{sh}$. It can be derived[2] from Eq. 10.1 that the minimum splitting resistance to an individual stud shear connector occurs when the width of the prism is

$$\left(b_c\right)_{min-one} = 3d_{sh} \tag{10.2}$$

and that the minimum splitting resistance is given by

$$\left(P_{min}\right)_{one} = 3.4d_{sh}^2 f_{cb}\left(\frac{27}{4} + \left[\left(1 - \frac{0.9d_{sh}}{\left(h_c\right)_{\leq 8.1d_{sh}}}\right)^2 \frac{0.9d_{sh}}{\left(h_c\right)_{\leq 8.1d_{sh}}}\right]^{-1}\right) \tag{10.3}$$

10.3.2   *Effective widths of prism*

The splitting resistances given in Eqs. 10.1 and 10.3 were derived for concentrically loaded prisms of width b_c. Eccentrically loaded prisms, such as those shown in Figure 10.4, can be analysed using an effective width concept that is often used in the design of the anchorage zone of post-tensioned members[1]. The effective width of the prism is the portion of the width of the eccentrically loaded prism that can be assumed to be loaded concentrically.

Examples of the effective widths of prisms are given in Figure 10.4 where the equivalent concentrically loaded prism of area $b_c \times h_c$ is shown shaded. Each stud shear connector acts concentrically on a prism of width b_c, where $b_c/2$ is the distance from the centre-line of the stud to the nearest free edge of slab or edge of the effective width of the slab as

s   h   o   w   n   .

It can be seen in (a) that stud A acts on a prism that has a greater effective width than stud B so that the splitting resistance to an individual stud connectors will governed by stud B. Similarly, in the composite L-beam in (b), the splitting resistance to stud D will be weaker than that to stud C, because the effective with of the former is smaller.

It is worth bearing in mind that it is the lateral dispersal of the load that causes splitting. Therefore, in the hybrid beam in Figure 10.4(d), the effective widths b_c are determined by the distance to the nearest horizontal edge and the effective height h_c is now the horizontal distance from the base of the stud that must not exceed $8.1d_{sh}$. It is also worth bearing in mind that the dowel force is concentrated at the base of the stud. Hence, dispersal of this force will cause splitting to start at the base of the stud, so that the effective width at the level of the base of the stud tends

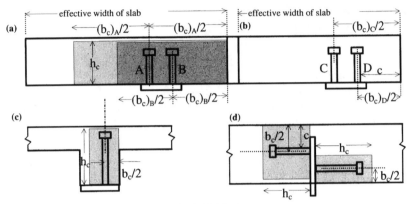

Figure 10.4 Effective prism widths for individual connectors

to control the splitting resistance as shown in the haunched beam in (c).

10.3.2.1 **Example 10.1** *Splitting resistance to an individual connector*

The stud shear connector in Example 5.2 has a diameter $d_{sh} = 19$ mm and a maximum dowel strength of $D_{max} = 100$ kN. The slab in which it is embedded has a depth of $h_c = 130$ mm and concrete strength of $f_c = 25$ N/mm², so that the tensile strength $f_{cb} = 0.5\sqrt{25} = 2.5$ N/mm². These stud shear connectors are to be used in a composite L-beam such as that in Figure 10.4(b) and it is necessary to determine the minimum effective width $(b_c)_D/2$ in (b) that will allow the dowel strength to be reached.

(a) Mean splitting resistance

As the depth of the slab $h_c = 130/19 = 6.8d_{sh} < 8.1d_{sh}$, the full depth of the slab can be used in Eq. 10.1 which becomes

$$P_{one} = 3069 \left[\frac{b_c}{19\left(1 - \dfrac{19}{b_c}\right)^2} + 10.08 \right] \quad kN$$

Furthermore from Eq. 10.3, $(P_{min})_{one} = 52$ kN when, from Eq. 10.2, $(b_c)_{min-one} = 57$ mm. These results are plotted in Figure 10.5 as the curve $(P_{one})_{3D}$; the suffix 3D refers to the fact that the concentrated load is being dispersed in three-dimensions. It can be seen that the intercept with $D_{max} = 100$ kN occurs at $b_c \approx 390$ mm. Hence, $(b_c)_D/2 \approx 195$ mm in Figure 10.4(b), so that the side cover required to the shank is $c \approx (390 - 19)/2 \approx 185$ mm.

(b) Characteristic splitting resistance

Splitting of plain concrete is a brittle mechanism that can lead to rapid failure if an alternative mechanism is not present to transfer the transverse forces after splitting. If it is necessary to prevent splitting, then it may be advisable to base the design on the 5% characteristic strength that tests have shown occurs at 76% of the mean values given in Eqs. 10.1 and 10.3. These results are plotted as $(P_{one})_{char}$ in Figure 10.5, from which $b_c \approx 590$ mm, that is a side cover of $c \approx 285$ mm would be required to prevent splitting. However if transverse reinforcement is supplied across the splitting zone as stipulated in Chapter 11, then failure can be considered to be ductile so that design could be based on a value close to the mean strengths[2] as in the previous calculation.

(c) Dispersal of concentrated load

The transverse component of the splitting resistance in Eq. 10.1 is shown in Figure 10.5 as $(P_{one})_{2D}$; where the suffix 2D refers to the fact that the concentrated load is being dispersed in only two directions. It can be seen that this is a lower bound to the three dimensional resistance $(P_{one})_{3D}$. It is also worth noting that the difference between $(P_{one})_{3D}$ and $(P_{one})_{2D}$ is constant and, therefore, vertical dispersal has its greatest benefit at low values of the effective width b_c.

(d) Minimum splitting resistance

The minimum splitting resistances shown in Figure 10.5, as given by P_{min} in Eq. 10.3, occur at an effective width of prism of 57 mm, as given by $(b_c)_{min}$ in Eq. 10.2. For effective widths less than $(b_c)_{min} = 57$ mm in Figure 10.5, the splitting resistance increases rapidly and is asymptotic to $b_c = d_{sh} = 19$ mm as shown. This increase in the splitting resistance at $b_c < (b_c)_{min}$ should be ignored in practice, so that when $b_c < (b_c)_{min}$, then it should be assumed that the splitting resistance $P_{split} = P_{min}$ as shown in Figure 10.5. This limitation is unlikely to occur with individual connectors because it would require a side cover of ·less than the diameter of the stud. However, it is worth noting that it can apply when dealing with groups of blocks of connectors as discussed in Section 10.4.

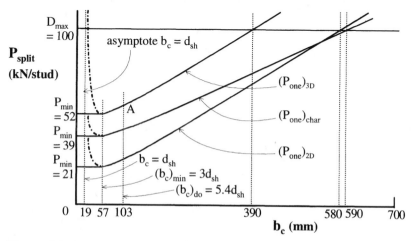

Figure 10.5 Splitting resistance of slab to individual connectors

(e) Minimum side cover to achieve the dowel strength

Stud shear connectors require a minimum side cover of

$$c_{do} = 2.2d_{sh} \tag{10.4}$$

to achieve the triaxial restraint to the concrete that is necessary to attain the maximum dowel strength D_{max}. The effective width from this cover is shown as $(b_c)_{do} = 5.4d_{sh} = 103$ mm in Figure 10.5. Effective widths less than $(b_c)_{do}$ should not be used in practice as crushing of the concrete may precede the splitting resistance. For example, the splitting resistance $(P_{one})_{3D}$ in Figure 10.5 intercepts $(b_c)_{do} = 5.4\,d_{sh} = 103$ mm, at point A at $P_{split} = 56$ kN. Hence, if it is not required to achieve the maximum dowel strength $D_{max} = 100$ kN, then the side cover to the stud can be reduced to 2.2 d_{sh}, that is $b_c = 5.4\,d_{sh}$, when splitting will occur at a load of 56 kN prior to dowel failure.

10.4 Effective dimensions for groups of connectors

Stud shear connectors can often be grouped together as in Figure 10.6(a). In this example the concrete prism is subjected to one lateral row of connectors that is positioned in three longitudinal lines as shown. The prism can split due to the concentrated load from an individual connector or from the combined effect of a group of connectors. It is, therefore, necessary to check for the splitting resistance to all possible combinations, however, engineering judgement can be used to minimize the number of checks required.

Each individual connector of the n connectors in the group in Figure 10.6(a) can cause the slab to split but as the connector in line 1 has the smallest effective width, the resistance of the concrete prism to this individual connector will be the least. Hence, it is only necessary check the splitting resistance to this individual connector using the procedure described in Section 10.3.1. It is worth noting that this individual connector acts as a concentric patch load of width $(b_a)_1 = d_{sh}$ on the prism of width $(b_c)_1$. Furthermore, the pair of connectors in lines 1 and 2 in Figure 10.6(a) act as a concentric patch load of width $(b_a)_p = t_p + d_{sh}$ on the prism of width $(b_c)_p$, where $(b_c)_p/2$ is measured from the centreline of the patch to the nearest side or effective side. Similarly, the group of n connectors act as a patch of width $(b_a)_n = t_n + d_{sh}$ on the prism of width $(b_c)_n$, where $(b_c)_n/2$ is measured from the centre of the patch to the nearest effective side.

10.5 Pairs of connectors

10.5.1   *Splitting resistance to pairs of connectors*

The mean splitting resistance of a concrete prism to a pair of connectors is given by

$$P_{pair} = 3.4\left(t_p + d_{sh}\right)d_{sh}f_{cb}\left(\frac{b_c}{\left(t_p + d_{sh}\right)\left(1 - \dfrac{t_p + d_{sh}}{b_c}\right)} + \left[\left(1 - \frac{0.9d_{sh}}{\left(h_c\right)_{\le 8.1d_{sh}}}\right)^2\frac{0.9d_{sh}}{\left(h_c\right)_{\le 8.1d_{sh}}}\right]^{-1}\right) \tag{10.5}$$

where the minimum splitting resistance is given by

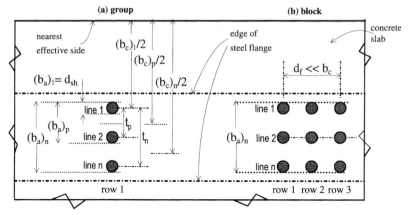

Figure 10.6 Effective dimensions for combinations of connectors

$$\left(P_{min}\right)_{pair} = 3.4\left(t_p + d_{sh}\right)d_{sh}f_{cb}\left(4 + \left[\left(1 - \frac{0.9d_{sh}}{\left(h_c\right)_{\leq 8.1d_{sh}}}\right)^2 \frac{0.9d_{sh}}{\left(h_c\right)_{\leq 8.1d_{sh}}}\right]^{-1}\right) \tag{10.6}$$

and which occurs at an effective width of

$$(b_c)_{min-pair} = 2(t_p + d_{sh}) \tag{10.7}$$

The main difference between Eqs. 10.5 and 10.1 is that the square in the denominator of the first term in Eq. 10.1 does not occur in Eq. 10.5. It is also worth noting that Eqs. 10.5 and 10.6 give the resistance of the slab to a force P imposed by a pair of connectors and, hence, these strengths P should be divided by two to determine the force per connector to cause splitting.

10.5.1.1 **Example 10.2** *Resistance to a pairs of connectors*
The connectors in the slab in Example 10.1 are to be placed in pairs. In order to be able to achieve the maximum dowel strength D_{max}, the connectors are spaced at the minimum lateral spacing of stud shear connectors, that is $L_T = 4d_{sh}$ as shown in Figure 5.3. Placing the pairs of connectors at this lateral spacing gives $t_p = 4d_{sh} = 76$ mm and applying this to Eq. 10.5 gives

$$\left(\left(P_{split}\right)_{t_p=4d_{sh}}\right)_{per\ stud} = 7671\left[\frac{b_c}{95\left(1 - \frac{95}{b_c}\right)} + 10.08\right]\ kN$$

which is plotted, in terms of the load per stud and side cover c in Figure 10.6(b), as line B in Figure 10.7. The minimum strength can be derived from Eq. 10.6 as $(P_{min})_{pair}$ = 108 kN/stud which exceeds the maximum dowel strength required of D_{max} = 100 kN so that there is no problem with splitting.

If the lateral spacing is reduced to $t_p = 2d_{sh}$, then Eq. 10.5 becomes

$$\left(\left(P_{split}\right)_{t_p=2d_{sh}}\right)_{per\ stud} = 4603\left[\frac{b_c}{57\left(1-\dfrac{57}{b_c}\right)}+10.08\right]\ kN$$

which is shown as line D in Figure 10.7 and, hence, a minimum side cover of 270 mm would be required to achieve a dowel strength of D_{max} = 100 kN.

10.6 Groups of connectors

10.6.1 *Splitting resistance to groups of connectors*

The mean splitting resistance of a concrete prism to the group of n connectors in Figure 10.6(a) is given by

$$P_{group}=3.4(t_n+d_{sh})d_{sh}f_{cb}\left(\frac{b_c}{(t_n+d_{sh})\left(1-\dfrac{t_n+d_{sh}}{b_c}\right)^2}+\left[\left(1-\frac{0.9d_{sh}}{(h_c)_{\le8.1d_{sh}}}\right)^2\frac{0.9d_{sh}}{(h_c)_{\le8.1d_{sh}}}\right]^{-1}\right) \quad (10.8)$$

which has a minimum splitting strength of

$$\left(P_{min}\right)_{group}=3.4(t_n+d_{sh})d_{sh}f_{cb}\left(\frac{27}{4}+\left[\left(1-\frac{0.9d_{sh}}{(h_c)_{\le8.1d_{sh}}}\right)^2\frac{0.9d_{sh}}{(h_c)_{\le8.1d_{sh}}}\right]^{-1}\right) \quad (10.9)$$

when

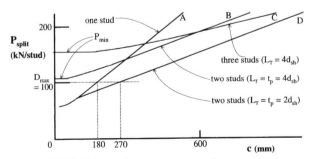

Figure 10.7 Splitting resistance to groups of connectors

$$\left(b_c\right)_{min-group} = 3\left(t_n + d_{sh}\right) \qquad (10.10)$$

10.6.1.1 **Example 10.3** *Resistance to groups of connectors*

Let us place the connectors in Example 10.1 in groups of three with the minimum recommended lateral spacing of $L_T = 4d_{sh}$ so that the spread of connectors, in Figure 10.6(a), $t_n = 152$ mm. Applying Eq. 10.8 gives

$$\left(\left(P_{split}\right)_{t_n=8d_{sh}}\right)_{per\ stud} = 9206\left[\frac{b_c}{171\left(1-\frac{171}{b_c}\right)^2} + 10.08\right]\ kN$$

that has a minimum splitting resistance of 154 kN/stud and which is plotted as line C in Figure 10.7. As with the two studs at $L_T = 4d_{sh}$ in Figure 10.7, the minimum splitting resistance of the group of three studs exceeds the dowel strength so that splitting will also not be a problem.

The results from the splitting resistance of individual connectors, from Figure 10.5, are also plotted in Figure 10.7 as line A. It can be seen that the splitting resistance to the individual connector controls the analysis for all the cases when the minimum recommended lateral spacing of $L_T = 4d_{sh}$ is adhered to.

10.7 Blocks of connectors

10.7.1 *Blocks of stud shear connectors*

Occasionally it may be necessary to concentrate the stud shear connectors in blocks as in Figure 10.6(b); this may be necessary in open composite truss girders or in composite stub girders. Let us define a block of connectors as several groups of connectors in which the longitudinal spread of the connectors d_f is much less than the effective width of the prism b_c as shown in (b). When d_f is of the same order of magnitude as b_c then we have the problem of global splitting which is dealt with elsewhere[2].

The block of connectors in Figure 10.6(b) can cause the slab to split through a single line of connectors and through combinations of these lines. In estimating the splitting resistance of the block of 9 connectors in (b), let us assume that the three lines shown are coincident with the three lines of connectors in (a), that is they are at the same distance from the effective side. As the effective width for the connector in line 1 in (a) is exactly the same as the effective width of the connectors in line 1 in (b), the splitting resistance of the concrete prism will be exactly the same. The same can be said for the pair of connectors in (a) and (b) and the group of n connectors in (a) and (b).

10.7.1.1 **Example 10.4** *Resistance to a block of connectors*

The stud shear connectors in Example 10.1 are to be used in an open truss girder where they are to be placed in blocks of 9 connectors as shown in Figure 10.6(b). The lateral spacing between individual connectors is to be maintained at $L_T = 4d_{sh}$. The splitting resistances to one row of this group as in (a) have already been determined and the results are given in Figure 10.7 (where the result for two studs at $L_T = 2d_{sh}$ is not applicable to this analysis). As there are three rows of studs in Figure 10.6(b), the splitting resistance per stud is one-third of the splitting resistance per stud in (a). In other words, the splitting resistance per stud for the block of connectors in (b) is one-third of the strengths given in Figure 10.7 and which are shown in Figure 10.8.

It can be seen in Figure 10.8 that the splitting resistance to all 9 connectors, that is the curve labelled '3 lines and 3 rows', now controls the strength and a side cover of 1660 mm is required to allow the total dowel force of 900 kN to be applied to the slab. It is worth comparing this analysis with that in Figure 10.7 where the strength of the individual connector controlled the design for all the cases except where $L_T = 2d_{sh}$.

10.8 Prisms with non-rectangular cross-sections

The previous sections dealt with splitting of rectangular prisms such as those shown in Figure 10.4. The procedure is now applied to prisms that have sloping sides and haunches.

10.8.1   *Upper and lower bound solutions*

Consider the non-rectangular prisms in Figure 10.9 in which the lower width is b_i, the upper width is b_o, the height of the prism $h_c \leq 8.1d_{sh}$, and the sides slope at an angle of θ degrees to the vertical.

The non-rectangular prisms in Figure 10.9 can be considered to be bound by a pseudo inner rectangular prism of width $b_c = b_i$, that is shaded, and a pseudo outer rectangular prism of width $b_c = b_o$, that is hatched. As the cross-section of the pseudo inner rectangular prism falls within the perimeter of the non-rectangular prism, the splitting resistance of the inner prism $(P_{split})_i$ will give a lower bound to the splitting resistance of the non-rectangular prism. Conversely, as part of the outer pseudo rectangular prism falls outside the non-rectangular prism, the strength of the pseudo outer rectangular prism $(P_{split})_o$ will form an upper bound to the strength. These upper and lower bounds can be determined from the splitting resistance of rectangular prisms already described.

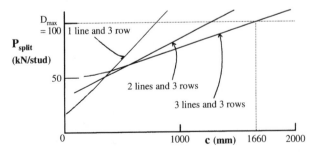

Figure 10.8 Splitting resistance to blocks of connectors

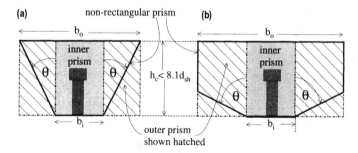

Figure 10.9 Standard non-rectangular sections

It is also worth noting that when $\theta = 0$ in Figure 10.9, the inner rectangular prism gives the correct strength and when $\theta = 90°$ the outer rectangular prism gives the correct strength. Tests have shown that the splitting resistance of the non-rectangular prism $(P_{split})_\theta$ is given by the following linear interpolation between the strengths of the pseudo inner and outer prisms.

$$\left(P_{split}\right)_\theta = \left(P_{split}\right)_i + \left(\left(P_{split}\right)_o - \left(P_{split}\right)_i\right)\frac{\theta^o}{90} \qquad (10.11)$$

10.8.2 *Equivalent prism concept*

In order to determine the splitting resistance of a non-rectangular prism, it is necessary to simplify its shape to one of the standard non-rectangular prisms in Figure 10.9. For example, consider the haunched prism in Figure 10.10. The non-rectangular section can be represented by an equivalent prism that is created with effective sloping sides such as the one shown shaded. As this equivalent prism falls within the cross-section of the slab, the splitting resistance of the equivalent prism will be a lower bound to the true resistance.

The range of effective sides that can be chosen to create the equivalent prism in Figure 10.10 varies from $b_i = (b_c)_{min}$ when the effective width of the inner prism is the shaded equivalent prism shown, to $b_i = b_h = b_o$ that is the equivalent prism is a rectangle of width b_h. It is a question of finding the effective side that gives the maximum splitting resistance. However even if the optimum side had not been chosen, the result would be a lower bound to the splitting resistance and, hence, safe. A simple design procedure would be to check the splitting resistances at the two extremities: when $b_i = (b_c)_{min}$ so that b_o is a function of the geometry of the slab and $(b_c)_{min}$; and when $b_i = b_o = b_h$.

The procedure used in Figure 10.10, where a range of effective sides can be used to determine the maximum splitting resistance, can also be applied to the haunched beam with sloping sides in Figure 10.11. Alternatively, the equivalent prism shown shaded, that follows the slope of the haunch, can be used to determine a lower bound to the splitting resistance.

The equivalent prism concept can also be used to determine the splitting resistance of composite beams with composite slabs, such as the composite slabs with dovetail ribs or trapezoidal ribs in Figure 10.12.

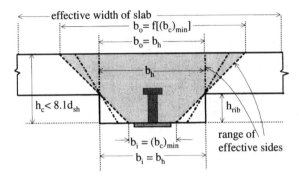

Figure 10.10 Effective sides

10.8.2.1 **Example 10.5** *Haunched beam with vertical sides*
(a) Single line

The stud shear connectors in Example 10.1 are placed in a single line in the haunched beam in Figure 10.13; the haunch has been inserted so that the height of the slab can be varied. It is required to find the variation of the strength of the shear connection with the depth of the rib of the haunch h_{rib}.

The side cover $c = 100$ mm, in Figure 10.13, exceeds $c_{do} = 2.2d_{sh} = 42$ mm (Eq. 10.4), hence, the dowel strength $D_{max} = 100$ kN can be achieved when splitting is prevented. We will use the equivalent prism shown shaded in Figure 10.13 to derive at least a lower bound to the strength. The equivalent prism has been chosen so that the inner prism has the minimum allowable splitting width of $3d_{sh}$ (Eq. 10.2) and, furthermore, the sides touch the top edge of the haunch as shown. Equations 10.1 and 10.3 are subjected to an upper bound of $h_c = 8.1d_{sh} = 154$ mm, hence, we will assume that $h_{rib} ³ 24$ mm so that $h_c ³ 154$ mm. Hence, for $h_c = 154$ mm, from Eq. 10.3, $(P_{split})_i = 56$ kN and from Eq. 10.1

$$P_o = 3069 \left[\frac{b_o}{19 \left(1 - \frac{19}{b_o}\right)^2} + 11.39 \right] \quad kN$$

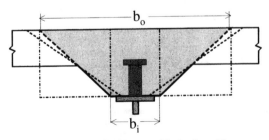

Figure 10.11 Haunched beam with sloping sides

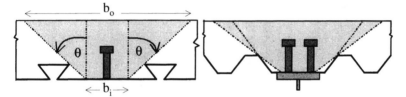

Figure 10.12 Effective sides of prisms in composite slabs

where from the geometry of the equivalent prism $b_o = (2 \times 154 \times 81)/h_{rib} + 57$
$= (24,948/h_{rib}) + 57$ mm.

Applying Eq. 10.11 for various values of h_{rib} gives the variation in strength in Figure 10.14 which is asymptotic to the minimum splitting resistance of 56 kN. It can also be seen in Figure 10.14 that the dowel strength of 100 kN can be achieved when $h_{rib} < 56$ mm and, therefore, for any larger values of h_{rib} the shear connector strength is reduced to that of the splitting resistance.

(b) Double line
Let us replace the single line of studs in Figure 10.13 with a pair of studs that has a lateral spacing $L_T = 4d_{sh} = 76$ mm. Furthermore, let us increase the width of the haunch by 76 mm, to 295 mm, so that the side cover to the studs remains at 100 mm. The splitting resistance to an individual connector remains unchanged as that shown in Figure 10.14. However, it is now necessary to check for the splitting resistance to the pair of connectors. The weakest prism is probably a rectangle with a width equal to that of the haunch of 295 mm. Applying Eq. 10.5 gives the splitting resistance per stud of $7671 \times [4.58 + 11.39] = 123$ kN, that exceeds the dowel strength as shown in Figure 10.14 and, hence, splitting induced by the pair of connectors does not affect the strength of the shear connection.

10.8.2.2 **Example 10.6** *Haunched beam with sloping sides*
The shear connectors in Example 10.1 are to be placed in a haunch of depth 150 mm as shown in Figure 10.15. It is necessary to make the base of the haunch as narrow as possible so that the side cover has been reduced to $c_{do} = 2.2d_{sh} = 42$ mm. It is required to determine the slope of the haunch that will allow the dowel strength of the shear connection to be achieved.

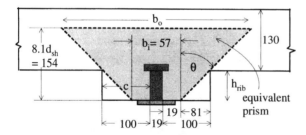

Figure 10.13 Varying depth of haunch

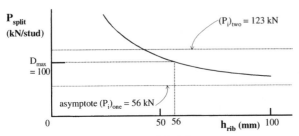

Figure 10.14 Variation in haunch height

An equivalent prism that follows the sides of the haunch has been chosen as shown in Figure 10.15. The splitting resistance of the inner prism $P_i = 3069 [8.15 + 11.39] = 60$ kN and that of the outer is given by the equation in Example 10.5 where, from the geometry of the haunch, $b_o = 103 + (2\times154) \tan\theta$ mm. The results are plotted in Figure 10.16 where it can be seen that it is necessary for the slope of the haunch to be greater than 55° for the dowel strength of 100 kN to be achieved.

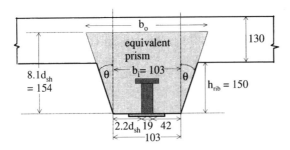

Figure 10.15 Haunched beam with sloping sides

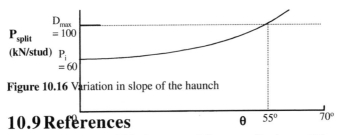

Figure 10.16 Variation in slope of the haunch

10.9 References

1. Leonhardt, F., (1964). Prestressed Concrete, Design and Construction. 2nd edn, Wilhelm Ernst and sohn, Berlin-Munich.
2. Oehlers, D. J. and Bradford, M. A. (1995). Composite Steel and Concrete Structural Members: Fundamental Behaviour. Pergamon Press, Oxford.

11 Post cracking dowel strength

11.1 Introduction

The stud shear connector of shank diameter d_{sh} = 19 mm in Example 5.2 has a maximum dowel strength of D_{max} = 100 kN when encased in concrete of compressive strength f_c = 25 N/mm². As can be seen in Figure 5.1, the stud bears against the concrete in a small zone that is adjacent to the base of the stud. If we assume that the area of the bearing zone is $d_{sh} \times d_{sh}$, then the mean bearing stress is 277 N/mm² = $11f_c$, that is considerably greater than the cylinder compressive strength of the concrete. The concrete in the bearing zone can only sustain this magnitude of stress because it is restrained triaxially by the stud and the adjacent flange. It can be seen that if the triaxial restraint is reduced, with the consequential reduction in the triaxial compressive strength, then the dowel strength of the stud shear connection will also reduce.

A longitudinal crack that extends through the bearing zone in Figure 5.1, that may have been induced by transverse positive moments in the slab of a composite beam or by splitting as described in Chapter 10, can reduce the triaxial restraint in the bearing zone and, hence, reduce the post-cracking dowel strength. For example, in the absence of any transverse reinforcement bridging the splitting crack in Figure 10.1(b), the concrete elements on either side will separate so that the post-cracking dowel strength D_{crack} = 0. However, tests have shown that judicious placement of transverse reinforcement across the crack plane can substantially increase the post-cracking dowel strength and, if required, allow the maximum dowel strength D_{max} to be achieved.

The effect of transverse reinforcement on the post-cracking dowel strength of stud shear connectors is dealt with in this chapter. The transverse reinforcement has the following very important functions after longitudinal cracking has occurred in the vicinity of the stud shear connectors:

- The transverse reinforcement maintains equilibrium by resisting the transverse tensile force $(F_t)_{ten}$ in Figure 10.2, that were originally resisted by the uncracked concrete; this is covered in Section 11.4.
- Confines the concrete in the bearing zone in Figure 5.1 so that the dowel action can be maintained and which is covered in Sections 11.2 and 11.3.
- Inhibits or arrests the propagation of the longitudinal crack[1].
- Allows a ductile mode of failure after splitting.

11.2 Hooped reinforcing bars

Composite L-beams such as that shown in Figure 10.4(b) are prone to longitudinal splitting along the line of the outer stud D. When the cover c is very small it may be impractical if not impossible to anchor straight reinforcing across the crack plane. This

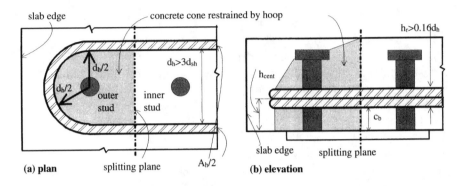

Figure 11.1 Hooped reinforcement

may be overcome by bending the transverse reinforcement around the outer stud as shown in Figure 11.1 where the transverse reinforcement has the dual role of confining the concrete and transferring the longitudinal shear into the concrete slab. This form of construction can be used in precast slabs for composite beams where pockets are left in the precast slabs at the positions of the shear connectors. After placing the precast slab on the steel beam, the pockets can then be grouted with a much stronger concrete than the slab, to enhance the dowel strength of the shear connection.

11.2.1   Dowel strength of studs with hooped reinforcement

Figure 11.1 illustrates a pair of studs that are close to the side of a composite L-beam and in which splitting has occurred along the longitudinal plane indicated. Tests have shown that the hooped reinforcement confines a cone of concrete around the outer stud as shown shaded, so that the longitudinal shear is transferred by dowel action into the reinforced concrete cone and thence by longitudinal shear action into the concrete slab. It is worth noting that the dowel strength of the inner stud is not affected by the longitudinal crack and, hence, can be assumed to be unchanged at D_{max}.

It has been explained previously using Figure 5.1 that the dowel strength of a stud shear connector depends on the eccentricity e of the resultant force F_1; the smaller the eccentricity the larger the dowel strength. The position of the resultant force F_1 in the outer stud in Figure 11.1 depends on the height of the transverse reinforcement h_{cent} and, therefore, the strength of this outer stud after splitting, D_{crack}, is a function of h_{cent} as shown in the following empirically derived equation.

$$D_{crack} = D_{max}\left(1.6 - \frac{0.4 h_{cent}}{d_{sh}}\right) \tag{11.1}$$

where h_{cent} is measured from the centroid of the transverse reinforcement to the base of the stud and in which $D_{crack} \le D_{max}$.

To achieve the dowel strength D_{crack} in Eq. 11.1, it is necessary to provide sufficient triaxial restraint to the concrete and this can be achieved by ensuring that

$$A_h \geq \frac{0.9d_h^2}{n}\left(\frac{D_{crack}}{D_{max}} - 0.4\right) \tag{11.2}$$

where A_h is the cross-sectional area of both arms of the hoop as shown in Figure 11.1(a) and n is the modular ratio E_s/E_c. Equation 11.2 applies when

$$d_h \geq 3d_{sh} \tag{11.3}$$

where d_h is the internal diameter of the bend of the hoop. Triaxial restraint is also maintained by the following equations that ensure that the hooped reinforcement does not yield and the concrete does not crush within the bend of the hoop.

$$A_h f_{yr} \geq 0.24d_h^2 f_c \tag{11.4}$$

$$h_r \geq 0.16d_h \tag{11.5}$$

where f_{yr} is the yield strength of the hooped reinforcement, that has to be fully anchored within the slab, and h_r is the depth of the hooped reinforcement as shown in Figure 11.1.

11.2.1.1 **Example 11.1** *Hooped transverse reinforcement*
The stud shear connectors in Example 5.1 are to be placed in a composite L-beam such as in Figure 10.4(b) where the cover c to the outer connector is 100 mm. The stud shear connection has the following properties: $D_{max} = 113$ kN; $f_c = 25$ N/mm^2; $f_{cb} = 2.5$ N/mm^2; $E_c = 25$ kN/mm^2; $E_s = 200$ kN/mm^2; hence n = 8; and $f_{yr} = 400$ N/mm^2.

From Figure 10.5, the splitting resistance of the slab to the outer connector is $P_{split} = 56$ kN which is considerably less than the required dowel strength of 113 kN. As there is very little room to anchor the transverse reinforcement across the splitting plane, hooped reinforcement will be used. It is worth noting that as splitting may occur at serviceability loads, as $P_{split} \ll D_{max}$, it may still be necessary to place nominal reinforcement across the splitting plane just to hold the concrete in place after splitting.

From Eq. 11.3, we will choose $d_h = 4d_{sh} = 76$ mm to allow for some construction tolerance. Inserting $D_{crack} = D_{max}$ into Eq. 11.2, as we require the maximum dowel strength, and $d_h = 76$ mm and n = 8, gives $A_h/2 = 195$ mm^2 which converts to two 12 mm diameter bars or one 16 mm diameter bar. From Eq. 11.5, $h_r \geq 12.2$ mm so for all intents and purposes we can use either the two 12 mm diameter bars or the one 16 mm diameter bar. Inserting $D_{crack} = D_{max}$ into Eq. 11.1, as we require the maximum dowel strength, gives $h_{cent} \leq 1.5d_{sh} = 29$ mm. Hence, if the 16 mm bar is being used then the maximum cover $c_b = h_{cent} - h_r/2 = 29 - 16/2 = 21$ mm, whereas, if the two 12 mm bars are being used then $c_b = 29 - 12 = 17$ mm.

11.3 Post-cracking confinement of concrete

The following confinement rules were determined empirically from tests on the dowel strength of stud shear connectors in longitudinally split slabs in which straight transverse reinforcement crossed the cracked plane. An example of the configuration of these tests is shown in Figure 11.2 where the reinforcing bars are transverse to the direction of thrust from the connectors. In these tests the main purpose of the transverse reinforcement was to confine the concrete in order to enhance the dowel strength, however, it is worth noting that the transverse reinforcement also resisted the transverse forces after splitting had occurred. Hence, the transverse reinforcement had the dual role of confining the concrete and resisting the transverse forces.

11.3.1 Dowel strength with straight transverse bars

The post-cracking dowel strength of stud shear connectors in longitudinally cracked slabs with straight transverse reinforcing bars is given by the following equation

$$D_{crack} = D_{max}\left(0.60 + \frac{0.93 A_b L_{con}}{d_{sh}^2}\right) \tag{11.6}$$

where L_{con} is the longitudinal spacing of the connectors as shown in Figure 11.2, A_b is the area of the bottom transverse reinforcement per unit length, $D_{crack} \le D_{max}$, and in which the characteristic strength, below which 5% of the results fall, can be obtained by replacing the coefficient 0.6 with 0.55. Equation 11.6 can be applied when A_b exceeds the following minimum requirement

$$A_b \ge \frac{0.065 d_{sh}^2}{L_{con}} \tag{11.7}$$

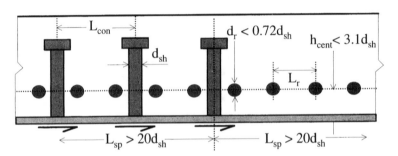

Figure 11.2 Transverse confinement reinforcement

It is worth noting that Eqs. 11.6 and 11.7 are dimensionally correct as A_b is the area per unit length and hence has the unit of length.

The following detailing rules ensure that the transverse reinforcement adequately confines the concrete in the stud shear connection bearing zone.

$$d_r \leq 0.72 d_{sh} \tag{11.8}$$

$$h_{cent} \leq 3.1 d_{sh} \tag{11.9}$$

$$L_{sp} \geq 20 d_{sh} \tag{11.10}$$

where d_r is the diameter of the transverse reinforcement as shown in Figure 11.2, h_{cent} is the distance from the centroid of the transverse reinforcement to the base of the stud, and L_{sp} is the spread of the transverse reinforcement on either side of the concentrated load as shown in Figure 11.2.

It is worth noting that because the primary purpose of this transverse reinforcement is to confine the concrete, it does not need to be fully anchored. It is suggested that an anchorage length sufficient to achieve a stress of 250 N/mm² is adequate for confinement purposes.

11.3.1.1 **Example 11.2** *Individual connector concrete confinement*
Let us consider the composite beam in Example 5.1 where $D_{max} = 113\,kN$ and $d_{sh} = 19\,mm$.

(a) Design of a single line of connectors
From Example 5.1, $L_{con} = 238$ mm and we wish to place confinement reinforcement so that the post-cracking dowel strength $D_{crack} = D_{max} = 113$ kN. From Eq. 11.6, $A_b = (0.4 \times 19^2)/(0.93 \times 238) = 0.65\ mm^2/mm = 0.65\ mm$, which would require 8 mm bars at a longitudinal spacing of $L_r = (\pi \times 8^2/4)/0.65 = 77$ mm centres. From Eq. 11.7, the minimum requirement for $A_b = 0.065 \times 19^2/238 = 0.099$ mm which is less than the requirement of 0.65 mm. From Eq. 11.8, $d_r \leq 0.72 \times 19 = 14$ mm, so that the 8 mm bars can be used. From Eq. 11.9, $h_{cent} \leq 3.1 \times 19 = 59$ mm so that the 8 mm transverse reinforcement has to be placed within a distance of 57 mm from the base of the slab. From Eq. 11.10, $L_{sp} \geq 20 \times 19 = 380$ mm and as this is greater than $L_{con} = 238$, the 8 mm bars can be spread uniformly throughout the slab.

(b) Double line of connectors
When there is a double line of connectors, the longitudinal spacing is twice that in the previous section (a) so that $L_{con} = 2 \times 238 = 476$ mm. From Eq. 11.6 and in order to ensure that $D_{crack} = D_{max}$, $A_b = (0.4 \times 19^2)/(0.93 \times 476) = 0.33$ mm, which would require 8 mm bars at 154 mm centres which is simply double the spacing in the preceding analysis (a). Thus having two lines of connectors reduces the required amount of confining reinforcement, as each bar confines each line of connectors.

(c) Assessment of an existing structure

Let us now assume that the composite beam already exists, that the transverse reinforcement had been designed for the transfer of longitudinal shear as in Chapter 6, and that a longitudinal split has occurred in the beam. It is required to determine whether this split has weakened the composite beam.

In Example 6.5, the composite beam was designed with a double line of connectors and the amount of transverse reinforcement required for longitudinal shear was $A_b = 0.49$ mm. In part (b) above, the amount of transverse reinforcement required for confinement is 0.33 mm and as this is less than that already provided, the shear connectors should reach their maximum dowel strength even though splitting has occurred.

It can be deduced from Example 6.5 that if there were a single line of connectors, then the composite beam would have been designed for longitudinal shear with the minimum reinforcement requirement of $A_b = 0.363$ mm. This is less than the confinement requirement in part (a) of this example of 0.65 mm and, hence, the maximum dowel strength cannot be achieved. Substituting $A_b = 0.363$ into Eq. 11.6 gives $D_{crack} = 0.60 + (0.93{\times}0.363{\times}238/19^2) = 0.82 D_{max} = 0.82{\times}113 = 93$ kN. Hence, splitting has reduced the strength of the shear connection by 18%, so that the strength of the composite beam will have to be reassessed using the partial shear connection procedures in Chapter 4.

11.3.1.2 **Example 11.3** *Confinement of a block of connectors*

Let us now design the confinement steel for the block of nine 19 mm diameter connectors in Example 10.4 which are shown in Figure 10.6(b), so that the post-cracking dowel strength D_{crack} is equal to the maximum dowel strength D_{max}. It will be assumed that the connectors are spaced at the recommended minimum longitudinal spacing of $L_L = 5d_{sh}$ in Figure 5.3(c). Hence $L_{con} = 5d_{sh} = 95$ mm. Applying Eq. 11.6 gives $A_b = (0.4{\times}19^2)/(0.93{\times}95) = 1.63$ mm which is much larger than the minimum requirement of $0.065{\times}19^2/95 = 0.25$ mm from Eq. 11.7; furthermore from Eq. 11.8, $d_r \leq 14$ mm. Hence we could use 10 mm bars at $L_r = (\pi{\times}10^2/4)/1.63 = 48$ mm within the group of connectors in Figure 10.6(b) and 12 mm bars at $L_r = 69$ mm extending either side of the group by $L_{sp} = 380$ mm. The bars should be placed within a distance of $h_{cent} = 59$ mm from the soffit of the slab.

11.4 Post-splitting transverse forces

Section 11.3 dealt with the transverse reinforcement required to confine the concrete in the vicinity of the shear connectors, in order to achieve either the maximum dowel strength or at least a reasonable dowel strength. After splitting, the concentrated load has still to be dispersed into the concrete slab and this can be achieved if transverse reinforcement is available to transmit the transverse forces.

It can be seen in Figure 10.2(a) and (c) that the concentrated load induces both transverse tensile forces and transverse compressive forces of equal magnitude. Hence, the resultant force across a splitting plane is zero which means that the transverse forces do not affect the global behaviour of the slab. Therefore, transverse flexural reinforcement that crosses the splitting plane can contribute to resisting the transverse splitting force $(F_t)_{ten}$.

An alternative way of viewing the problem is shown in Figure 11.3 where the local stress resultants are shown in a region of the slab where there is a local transverse tensile force F_t due to splitting, such as in the slab below the patch load in Figure 10.2(c). Let us assume that the transverse hogging moment applied to the slab of the composite beam has yielded the reinforcement so that transverse tension due to flexure across the cracked plane is $F_{ten} = F_f = A_t f_{yr}$ as shown in Figure 11.3. Hence for equilibrium along the cracked plane, the transverse compressive force $F_{cmp} = F_f - F_t$. In order to simplify the illustration of this mechanism we will assume that F_t is in line with F_{cmp}.

Let us start by assuming that the slab in Figure 11.3 is not subjected to splitting forces but is only resisting the full flexural capacity so that $F_t = 0$ and, hence, $F_{ten} = F_{cmp} = F_f = A_t f_{yr}$. As the splitting force F_t is increased, F_{ten} remains the same as the top steel is fully yielded so that the compressive force F_{cmp} reduces. However, the resultant force at the bottom of the slab $F_{cmp} - F_t$ remains unchanged at $F_f = A_t f_{yr}$, so that the moment capacity is being maintained. It can be seen that the transverse tensile force has not affected the moment capacity, which means that the transverse reinforcing bars provided for flexure also act in resisting splitting and which also means that additional reinforcement is not required. This situation remains stable whilst $F_t \leq F_f = A_t f_{yr}$. When $F_t > A_t f_{yr}$ then additional reinforcement will have to be provided so that the strength of the transverse reinforcement is now governed by the splitting force F_t.

It is worth noting that the mechanism illustrated in Figure 11.3 requires that the top flexural reinforcement is ductile, as it first extends to resist flexure and then further extends to resist the splitting forces. If there are problems with the ductility of the reinforcement, then it may be necessary to add more reinforcement.

11.4.1   Transverse splitting forces

A conservative estimate of the tensile splitting force T in Figure 10.1 is to assume that none of the concentrated load is dispersed vertically as shown in (a) but all of

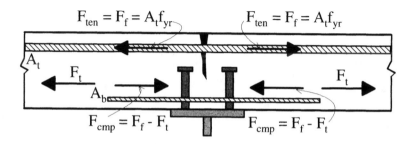

Figure 11.3 Local distribution of stress resultants

the concentrated load is dispersed longitudinally and transversely as in (b). For this case the transverse tensile force F_t, which is shown as T in Figure 10.1, is given by

$$F_t = \frac{P_{split}}{\pi}\left(1 - \frac{b_a}{b_c}\right)^2 \tag{11.11}$$

where P_{split} is the concentrated force; and where for individual lines of connectors, such as those shown in Figure 10.6, $b_a = d_{sh}$; for two adjacent lines of connectors $b_a = (t_p + d_{sh})$ and the exponent 2 should be replaced by 1; and for n lines of connectors $b_a = (t_n + d_{sh})$.

It is necessary to ensure that the transverse reinforcement that is provided to resist F_t is fully anchored. The distribution of the transverse tensile stresses prior to splitting is shown in Figure 10.2, where it can be seen that the longitudinal spread is a function of the effective width b_c and that it is concentrated near the patch load. Hence, it is suggested that the transverse reinforcement required to resist F_t is spread in front of the connector, that is in front of the thrust, and over a length of approximately $0.5b_c$ from the connectors. It is still necessary to supply transverse reinforcement in front of and behind the connectors to confine the concrete.

11.4.1.1 **Example 11.4** *Splitting force for block connectors*
In the block of 9 stud shear connectors in Example 10.4 (that have already been analysed for confinement in Example 11.3), $P_{split} = 900$ kN, $b_a = 9d_{sh} = 171$ mm, c = 1660 mm and, hence, $b_c = 2\times1660 + 171 = 3491$ mm. Inserting these values into Eq. 11.11 gives $F_t = (900/\pi)(1-171/3491)^2 = 259$ kN which has to be placed in the splitting zone of length $0.5b_c = 1.7$ m in front of the block; 8 mm bars of $f_{yr} = 400$ N/mm^2 at longitudinal spacing of $L_r = (1700/259\times10^3)\times(\pi 8^2/4)\times400 = 132$ mm will suffice.

11.4.1.2 **Example 11.5** *Splitting force for individual connector*
Let us continue the analysis in Example 11.2 where the stud shear connector force is $D_{max} = 113$ kN and where the transverse reinforcement for a single line of connectors was determined to be $A_b = 0.65$ mm and which was spread uniformly throughout the slab. As $L_{con} = 238$ mm, the strength of the confining reinforcement per stud is $F_t = 0.65 \times 236 \times 250 = 38$ kN (it has been assumed that the minimum bond strength has been used for this confinement reinforcement, that is it will allow a stress of 250 N/mm^2). It can be seen in Eq. 11.11 that an upper bound to the lateral tensile force is $P_{split}/\pi = D_{max}/\pi = 113/\pi = 36$ kN/stud which is less than the strength of the confining reinforcement of 39 kN/stud and, hence, extra reinforcement is not required.

11.5 Reference
1. Oehlers, D. J. and Bradford, M. A. (1995). Composite Steel and Concrete Structural Members: Fundamental Behaviour. Pergamon Press, Oxford.

12 Rigid plastic analysis of continuous composite beams

12.1 Introduction

The previous chapters have presented enough material to enable a static analysis of simply supported composite beams to be carried out, either with the elastic assumptions of Chapter 3 or the rigid plastic assumptions of Chapter 4. The analysis based on either assumption is straightforward, since simply supported beams subjected to gravity loads experience sagging bending throughout and are statically determinate. In the elastic analysis of these beams with full shear connection, we may use the full-interaction flexural rigidity based on the transformed cross-section $E_c I_{nc}$ for the whole length of beam. Furthermore, it was demonstrated in Chapter 4 that rigid plastic analyses of simply supported beams, even with partial shear connection, are not difficult. This chapter will extend the rigid plastic analyses of Chapter 4 to continuous composite beams, and will highlight that with certain conditions being met, the assumptions made in the rigid plastic model lead to a method of analysis that is extremely efficient and allows considerable increases in the loads that the beam can resist above those that are determined from an elastic analysis.

Continuous beams are statically indeterminate, and experience hogging or negative moments over the internal supports as well as sagging or positive moments within the spans. It was shown in Section 3.3.2 that the elastic analysis of these beams is not straightforward, as the points of contraflexure are not known in advance because of the different flexural rigidities in the sagging and hogging regions. An iterative solution is therefore required to solve for the redundant actions using either standard stiffness or flexibility methods of structural analysis. This difficulty can be overcome by use of a *rigid plastic analysis*, and the failure loads of a continuous composite beam may be determined quite readily by this method. Not only is rigid plastic analysis much easier than elastic analysis of continuous composite beams, but it is also more efficient as the shape factors, which are the ratio of the fully plastic moment to the moment to cause first yield, in both sagging and hogging are quite high, and this leads to an increase in failure loads above those determined from elastic analysis.

Rigid plastic analysis was initially developed some fifty years ago for mild steel beams, where it was demonstrated that the rigid plastic assumptions resulted in an increase in the failure load over that based on an elastic analysis with first yield, significantly so in some cases. The main assumption in a rigid plastic analysis is that the cross-section is *ductile*, and can experience large curvatures before failure, as was indicated in Section 2.3.3. It is perhaps logical to introduce the concepts of plastic analysis in continuous uniform *steel* beams first, and then to extend these concepts to composite beams, which of course have different moment capacities in sagging and hogging bending. There are also different ductility requirements for sagging bending of composite beams to those for steel beams, since the ductility of composite beams depends on the crushing of the concrete component. The following analysis presented in Section 12.2 for steel beams is merely to illustrate primarily the concepts of plastic hinges and moment redistribution. There is a proliferation of standard textbooks that treat plastic analysis and design in steel beams and frames fully (and are cited in Ref. 1), that include theorems and other aspects that are beyond the scope of this book. The reader should make recourse to these texts for a full understanding of the plastic analysis and behaviour of steel beams.

12.2 Continuous steel beams

12.2.1 *The plastic hinge*

Real moment-curvature response

Consider the mild steel I-section beam shown in Figure 12.1(a), where the actual stress-strain curve for the steel is shown in Figure 1.8, that is subjected to increasing curvature κ and which is free from residual stresses. We will make the usual engineering assumption that plane sections remain plane, so that the strain distribution varies linearly throughout the section as in (b). The response is initially elastic until the strain at the outermost fibre of the section reaches the yield strain $\varepsilon_y = f_y/E_s$ at the moment to cause first yield $M_y = f_y Z$, where Z is the elastic section modulus. As the curvature further increases, so too do the strains at the top and bottom of the section increase beyond the yield strain, so that yielding spreads down the section at f_y as in Figure 12.1(c). When the curvature produces a strain at the extreme fibres of the section equal to the strain hardening strain ε_{st}, most of the section has yielded and on increasing the curvature still further the beam enters the strain hardening region, so that the stresses increase in the highly strained regions of the section, as in (d). Theoretically increasing the curvature still further will finally produce failure when the strain reaches that to cause fracture in tension or inelastic local buckling in compression.

It is convenient to model the stress distribution in Figure 12.1(d) with the fully-plastic stress block shown in (e). Although the small elastic zone near the centroid is assumed to be fully yielded, this unconservative assumption is negated by ignoring

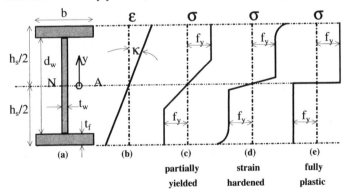

Figure 12.1 Strains and stresses on a bare steel I-section with increasing curvature

the real stresses in the outermost portions of the cross-section that are in the strain hardening zone and thus exceed f_y.

By increasing the curvature in steps, the bending moment may be obtained by taking the moment of the stresses over the area of the section, and which are of the shape depicted in Figures 12.1(c) and (d), producing the actual moment-curvature response shown in Figure 12.2. The moment of the idealized stress block shown in Figure 12.1(e) is of course the fully plastic moment $M_{ps} = f_y S_s$, where S_s is the plastic section modulus, and which is shown in Figure 12.2 along with M_y. It can be seen that the difference between the actual moment and the plastic moment at the strain hardening curvature κ_{st} is minuscule, since the elastic zone is very small and located close to the centroid of the section.

12.2.1.2 Idealized moment-curvature responses

In lieu of generating the real moment-curvature relationship which is arithmetically involved (although not significantly so), and because the real curve is not required to determine the collapse loads in the rigid plastic analysis, two idealizations of this real curve will be made for use subsequently. Firstly we can assume that the moment curvature response is linear elastic with flexural rigidity $E_s I$ until the plastic moment M_{ps} is attained, and is then constant at a plateau at M_{ps} as the curvature tends to infinity, as shown in Figure 12.2. Secondly, if the rigid plastic idealization of the rigid plastic stress-strain curve for the steel shown in Figure 1.13 is adopted, no elastic curvatures develop and the curvature is zero until M_{ps} is reached. The moment then remains constant at M_{ps} as shown in Figure 12.2. Although the rigid plastic idealization forms the basis of the analysis of continuous beams, the elastic-perfectly plastic moment curvature idealization will be used in Section 12.2.3 to demonstrate the concept of moment redistribution upon which plastic analysis is based.

12.2.1.3 *Plastic hinge in a steel member*

A simply supported steel beam is shown in Figure 12.3(a) subjected to a central

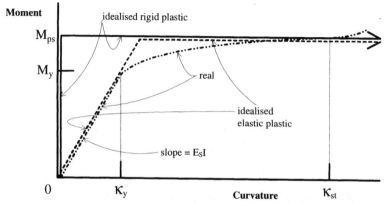

Figure 12.2 Real and idealized moment-curvature relationships for a bare steel I-section

concentrated load W. The bending moment diagram is known from statics and has a maximum value of WL/4, as shown in (b). If the load W is sufficiently large to produce moments that extend well into the inelastic range, the real moment-curvature response shown in Figure 12.2 may be used in conjunction with the bending moment diagram in Figure 12.3(b) to generate the curvature diagram that consists of elastic, plastic and strain hardened portions, as shown in (c). The curvature may be integrated to produce the variation of beam slope θ as in (d), and this may be in turn integrated to produce the variation of deflection v, as in (e).

Several related observations may be made from the real variations of curvature, rotation and deflection shown in Figures 12.3(c) to (e) respectively. Firstly it can be seen from (c) that when the beam is loaded well into the inelastic range the elastic curvatures are very small in comparison with the inelastic (plastic and strain hardened) curvatures. This is then reflected in the rotation of the beam in (d), which remains nearly constant in the elastic range at $\pm\theta$ on each side of the load as shown in (e), except adjacent to the load point where it varies very rapidly from $+\theta$ to $-\theta$ as in (d). Finally, as the slope is nearly constant except near the concentrated load, the distribution of deflection is close to linear (meaning that the beam deflects as a near to straight bar) except in the vicinity of the loading point.

The real behaviour shown in Figures 12.3(a) to (e) may be replicated closely if we adopt the rigid-plastic assumptions for the steel beam shown in Figure 12.2. Because the curvature in the real beam is small except close to the region of the concentrated load, this curvature is taken as zero and infinite at the load position, as shown in Figure 12.3(g). This assumption then leads to the rotations being constant at $\pm\theta$ either side of the load, as in (h), so that the beam deflects linearly as rigid bars on either side of the load point, as in (i). Under the load there is thus a 'kink' of angle 2θ which is shown in (i).

Because the curvature under the load is infinite, the corresponding moment is the maximum moment the beam can resist, M_{ps}, and the beam will fail when $WL/4 = M_{ps}$ or when $W_{collapse} = 4M_{ps}/L$. The plastic region shown in Figure 12.3(a) is now assumed concentrated directly under the load as shown in (f) and a hinge forms under the load. This *plastic hinge* is associated with the 'kink' under the load shown in (i). In this case, the rotation at the plastic hinge cannot be determined explicitly by the rigid-plastic idealization, but once the load reaches $W_{collapse}$ the beam is assumed to reach a mechanism condition. In this sense, the plastic hinge is analogous to a 'structural hinge', except that in the latter case the moment is zero while at a plastic hinge the moment is M_{ps}. It is worth reiterating that the underlying assumption of the plastic hinge concept is that the steel is ductile.

12.2.2 *Requirements for plastic analysis of steel beams*

The major requirement for plastic analysis of steel (or composite) beams is that plastic hinges can form, as will be illustrated in Section 12.2.3. This of course is a ramification of the ductility of the cross-section. Although as a material mild steel is ductile, the required rotations at plastic hinges may not be achieved due to premature failures associated with buckling. Local buckling may be prevented my ensuring that

Case 1: Real beam

(a) beam

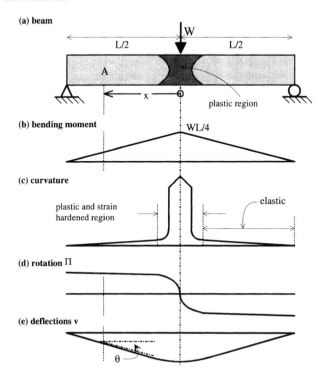

(b) bending moment

(c) curvature

(d) rotation Π

(e) deflections v

Case 2: Rigid plastic assumption

(f) beam

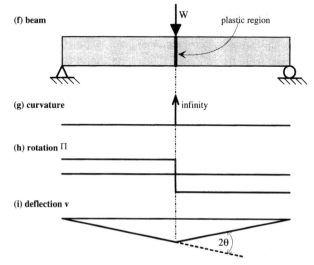

(g) curvature

(h) rotation Π

(i) deflection v

Figure 12.3 Plastic hinge concept

the section classification in Section 2.3.3 is at least *compact*. The geometric limits placed on the elements of a compact steel section ensure that inelastic local buckling will occur at large curvatures well into the inelastic range, and that the plastic moment is at least achievable before failure caused by inelastic local buckling. Statically indeterminate beams require a quantifiable amount of rotation at a plastic hinge to be achieved prior to inelastic local buckling occurring at the hinge, and such beams require the more stringent *plastic* classification (Section 2.3.3.2) which allows higher curvatures to be obtained.

The other requirement is that lateral-torsional buckling does not take place prior to collapse based on rigid-plastic analysis. This requirement is achieved by ensuring that the spacing of lateral restraints or braces satisfies specified limits in codes of practice[1]. However, Chapter 13 will demonstrate that the lateral-distortional buckling introduced in Section 1.6.2 that is associated with composite beams is far harder to quantify than lateral-torsional in plain steel members, but nevertheless must be prevented from occurring.

12.2.3 *Plastic analysis of continuous steel beams*

12.2.3.1 General

In continuous beams, at least two plastic hinges must form to produce a collapse mechanism in which the beam is able to deflect freely by rotating freely at the plastic hinges. Associated with the formation of this collapse mechanism is the concept of *moment redistribution*, that we will see in Section 12.3 is very important in composite beams. The analysis of a continuous steel beam will be carried out by means of a numerical example in Section 12.2.3.2, where the simplicity of calculating the collapse load will be shown ultimately.

12.2.3.2 **Example 12.1** *Analysis of two-span steel beam*

Consider a continuous steel beam whose dimensions are shown for the steel component in Figure 4.3 that is simply supported at the ends and over an internal support. Each span is of length 6 m, and subjected to concentrated loads at midspan of magnitude W. The symmetry of this problem allows us to analyse only one span, which must be modelled as a propped cantilever as shown in Figure 12.4(a). The properties of the cross-section are $I = 222.8 \times 10^6$ mm^4, $Z = 1.173 \times 10^6$ mm^3 and $S_s = 1.338 \times 10^6$ mm^3 with $E_s = 200$ kN/mm^2 and $f_y = 250$ N/mm^2.

The flexural rigidity of this uniform steel beam is constant along the length (unlike a composite beam) and the relevant moments in the elastic bending moment diagram shown in Figure 12.4(b) are given in structural engineering handbooks as $M_{hog} = -6WL/32$ and $M_{sag} = 5WL/32$. The maximum (elastic) moment is thus in the hogging region. The first yield moment is $M_y = 250 \times 1.173 \times 10^6$ Nmm $= 293.3$ kNm, so that first yield will occur when $W = (293.3 \times 32)/(6 \times 6) = 260.7$ kN. We will now use the elastic-perfectly plastic assumption shown in Figure 12.2 that assumes the beam is elastic until a plastic hinge forms, which will be at the internal support (modelled as the root of the propped cantilever shown as position A in Figure 12.4(a)). Hence using $M_{ps} = 250 \times 1.338 \times 10^6$ Nmm $= 334.5$ kNm, the load to cause a first hinge to form

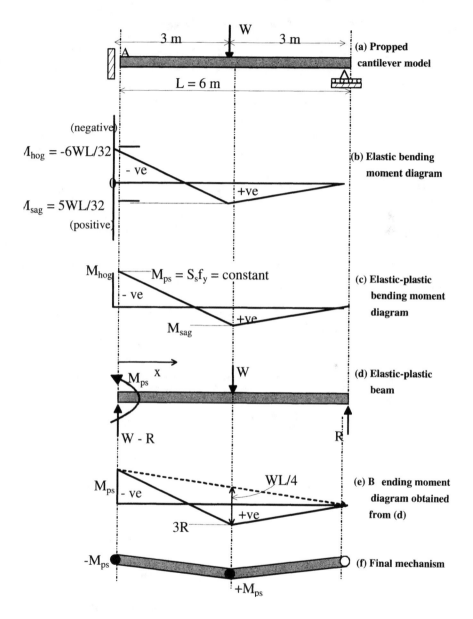

Figure 12.4 Two-span continuous beam

is W = (334.5 × 32)/(6 × 6) = 297.3 kN. Note that this is an increase of 297.3/260.7 or 14% above the first yield load. In addition, the moment under the load is $M_{sag} = 5 \times$ 297.3×6/32=287.7 kNm.

When a plastic hinge forms at the interior support (modelled as the root of the right hand span in Figure 12.4(a)), it is obvious that the span has not reached a collapse mechanism, but now resembles a simply supported beam (but with a moment of $-M_{ps}$ at the interior support). The beam can now be analysed when W is increased above its first hinge value of 297.3 kN by the elastic-perfectly plastic model assumed, and the bending moment diagram for this condition is shown in Figure 12.4(c). In this elastic analysis we must note that M_{hog} remains constant at $-M_{ps}$, and most importantly that the beam is free to rotate θ_{hog} at the interior support. Although the analysis is still elastic, the ramifications of the latter behaviour is that M_{sag} no longer equals 5WL/32, as this 'elastic solution' is derived on the assumption that the beam is built-in at the root, that is $\theta_{hog} = 0$. The elastic analysis must therefore be modified as follows to allow θ_{hog} to be non zero.

The beam is shown in Figure 12.4(d) when W > 297.3 kN, with the reaction at the right hand simple support being denoted R. The bending moment diagram corresponding to this state is shown in (e) where the moment at mid-span is 3×R (the reaction R times the lever arm of 3 m) and the static moment is WL/4. Clearly the distribution of moment M along the beam is given by

$$M = E_s I \frac{d^2 v}{dx^2} = (W - R)x - 334.5 - W\langle x - 3 \rangle \tag{12.1}$$

which upon successive integrations produces

$$E_s I \theta = (W - R)\frac{x^2}{2} - 334.5x - \frac{W}{2}\langle x - 3 \rangle^2 + a_1 \tag{12.2}$$

$$E_s I \ v = (W - R)\frac{x^3}{6} - 167.25x^2 - \frac{W}{6}\langle x - 3 \rangle^3 + a_1 x + a_2 \tag{12.3}$$

in which a_1 and a_2 are constants of integration, x is measured from the root of the propped cantilever as show, and $\langle \ \rangle$ represent Macaulay brackets[2], that is the term in the bracket is ignored when negative. The constants a_1 and a_2 may be determined by imposing the boundary conditions of zero deflection at the two ends, viz. v(0) = v(6) = 0, producing $a_2 = 0$ and

$$a_1 = 1004 + 6R - 5.25W \tag{12.4}$$

It is worth noting again that the condition $\theta(0) = 0$ does *not* hold as the beam is free to rotate at the plastic hinge at x = 0 with its moment remaining at $-M_{ps}$. If we sum moments about the internal support (or root of the propped cantilever idealization) then from equilibrium in Figure 12.4(d) $3W = 334.5 + 6R$. Substituting this into Eq. 12.4 and then into Eq. 12.2 and putting $x = 0$ produces the rotation at the hogging plastic hinge as $E_s I\theta(0) = E_s I\theta_{hog} = a_1$, or

$$\theta_{hog} = 50.5 \times 10^{-9}\left(297.3 - W\right) \quad \text{(radians)} \tag{12.5}$$

where W is in kN.

As W is increased above its first hinge value of 297.3 kN, a (negative) rotation θ_{hog} develops at the internal support and the moment under the concentrated load shown in Figure 12.4(e) increases until finally $M_{sag} = M_{ps}$ when $3R = 334.5$ kNm, or $R = 111.5$ kN. Therefore, as it has already been shown that $3W = 334.5 + 6R$ then $3W_{collapse} = 334.5 + (6 \times 111.5)$, that is $W_{collapse} = 334.5$ kN. The continuous beam (or propped cantilever model) has now been loaded until it forms a plastic mechanism as in (f) at a collapse load of $W_{collapse} = 334.5$ kN. In order to achieve this collapse mechanism, from Eq. 12.5 the hogging hinge has been required to rotate $50.5 \times 10^{-9} \times (297.3 - 334.5)$ or -1.88×10^{-6} radians. However, once $W_{collapse}$ has been attained, both the hogging and sagging hinges are free to rotate as a mechanism with an undetermined (but theoretically large) magnitude.

This example has illustrated a number of concepts that are unique to plastic analysis. Firstly, the load to cause plastic collapse of the continuous beam is 334.5/297.3 or 13% greater than that to produce a first hinge, so that the beam strength is not fully utilized when based on a first hinge analysis. Secondly, to reach failure a quantifiable rotation is required at the position of the hinge that forms first. Thirdly, and most importantly, the calculation of the collapse load is very easy. Although performed slightly differently in the previous calculations, it can be seen directly from the variation of the bending moment diagram in Figure 12.4(e) that at the position of the applied load W, that the static moment at collapse $= WL/4 = W_{collapse}L/4 = M_{ps} + M_{ps}/2$ producing $W_{collapse} = (4 \times 1.5 \times 334.5)/6 = 334.5$ kN, a calculation which is greatly simpler than that for elastic analysis of a statically indeterminate beam. Finally, it can be seen from the bending moment diagram in (e) that the shape of the bending moment distribution changes constantly as W is increased from its value to form a first hinge to that which causes collapse when a mechanism is reached. At the first hinge load, the sagging moment $M_{sag} = 287.7$ kN as noted earlier, but as the load is increased above its first hinge value, M_{hog} remains constant while moment is redistributed to the positive bending region, so that ultimately there is a moment redistribution 334.5/287.7 or 16% in the sagging region. This moment redistribution is a characteristic of plastic analysis, and is permitted by the ability of the plastic hinges to rotate freely due to their ductility.

12.3 Continuous composite beams

12.3.1 *General*

The presentation of plastic analysis in Section 12.2 for bare steel beams illustrated the simplicity and economy of design based on such analysis. Plastic analysis may also be used for composite beams, and the same benefits that are achieved in steel design also accrue to continuous composite beams. There are two major differences with composite beams compared with uniform steel beams. Firstly the plastic moments of resistance in composite beams are different in hogging and bending regions. The second difference relates to the ductility of the sagging region in particular, and this must be ensured by the requirements outlined in Section 12.3.2 so that the rigid-plastic assumptions of Chapter 4 that are used to calculate the plastic moment are valid.

12.3.2   *Composite plastic hinges*
12.3.2.1 Hogging behaviour
Under hogging bending, the real moment-curvature response is similar to that shown in Figure 12.2 for a steel beam, since the concrete cracks at low curvatures when the stress exceeds f_{ct} in Section 1.3.5.1, and the cross-section comprises of a 'steel' section consisting of the steel component and the reinforcement, as discussed in Chapter 3. In composite plastic analysis, we may make the usual rigid plastic assumptions unreservedly that were made in Section 12.2, provided of course that local buckling is prevented prior to a collapse mechanism developing, and that lateral-distortional buckling is also prevented. The local buckling provision is enforced by ensuring that the cross-section is compact or plastic, while the possibility of lateral-distortional buckling can be checked by the design models given in Chapter 13. The plastic hinge concept is thus the same as that for a bare steel cross-section, where at the plastic moment $(M_p)_{hog}$ the cross-section is allowed to rotate freely.

12.3.2.2 *Sagging behaviour*
The real moment-curvature relationship may be generated by modifying the model described for a bare steel section described in Section 12.2.1.1. However, under certain conditions the moment-curvature curve may reach the sagging plastic moment $(M_p)_{sag}$ and then decrease at relatively low curvatures owing to premature crushing of the concrete. Indeed, in some cases the value of $(M_p)_{sag}$ may not even be attained as the concrete stress-strain curve, as shown in Figure 1.10, is not ductile. Typical real moment-curvature responses of composite cross-sections subjected to a sagging moment are shown in Figure 12.5.

If the sagging hinge forms first, it must be ductile enough to allow sufficient rotation for the next hinge to form, as was quantified for the interior support region in Example 12.1, although this was a hogging hinge. In conventional reinforced concrete analysis[3], a singly reinforced beam is assumed to be ductile if the steel reinforcement yields before the concrete crushes at a strain ε_u. Based on an extensive parametric study, Rotter and Ansourian[4] proposed a similar ductility requirement for composite beams in sagging bending (that implicitly assumed full shear connection), except that the ductility requirement was that the bottom fibres of the steel component reach the strain hardening strain ε_{st} prior to crushing of the concrete component. This ductility requirement was modified for rigid plastic stress blocks in Ref. 5, and

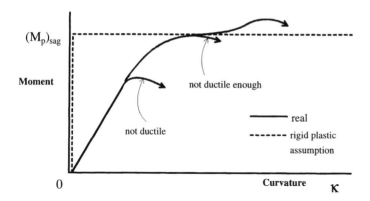

Figure 12.5 Moment-curvature response of a composite beam in sagging bending expressed as

$$\chi > 1 \tag{12.6}$$

in which the ductility parameter χ is expressed as

$$\chi = \frac{0.85 f_c b_c \varepsilon_u D}{A_s f_y (\varepsilon_u + \varepsilon_{st})} \tag{12.7}$$

where b_c is the width of the concrete component, D is the total depth of the composite beam that is $h_c + h_s$, and A_s is the cross-sectional area of the steel component.

The requirement of Eq. 12.6 has been found to be unconservative in situations where severe rotation is required at the sagging plastic hinge of a composite beam. Based on tests performed by Ansourian[6] on continuous beams which required substantial rotation capacity for the moments to redistribute and ultimately form a plastic mechanism, it has been suggested[5,6] that the ductility requirement

$$\chi > 1.6 \tag{12.8}$$

be satisfied. This allows for curvatures well into the strain hardening region, and allows the rigid plastic assumption of the curve shown in Figure 12.5 to be used.

12.3.2.3 **Example 12.2** *Calculation of ductility parameter*

The cross-section shown in Figure 4.3(a) is assumed to have a strain hardening strain $\varepsilon_{st} = 11\varepsilon_y = (11 \times 250)/(200 \times 10^3) = 0.01375$ and the ultimate strain ε_u in the concrete is taken as[5] 0.0033. The area of the steel component is $A_s = 9200$ mm². Hence from Eq. 12.7, $\chi = (0.85 \times 25 \times 3500 \times 0.0033 \times (130 + 380))/(9200 \times 250 \times (0.0033 + 0.01375)) = 3.2 > 1.6$, and this cross-section may thus undergo large rotations at the plastic hinge in order for other hinges to develop and lead to a collapse mechanism.

12.3.3 *Plastic analysis of continuous composite beams*

12.3.3.1 General

The advantages of plastic analysis in bare steel continuous beams were made obvious in Section 12.2. These advantages are even greater for continuous composite beams, and follow the same arguments that were presented in Section 12.2. In the latter section, an elastic analysis was used to calculate the first-hinge collapse load, but it was demonstrated in Chapter 3 that such an elastic analysis is difficult for a continuous composite beam. Since we seek to calculate the *collapse load*, a first-hinge analysis is unnecessary as it does not produce a mechanism condition in the beam. Hence only the final collapse mechanism need be chosen, and with the values of $(M_p)_{sag}$ and $(M_p)_{hog}$ which are easily calculated for a cross-section, determining the collapse load is usually straightforward. It is worth noting again that the basis of the method of plastic analysis is the rigid plastic analyses of a beam with full shear connection discussed in Chapter 4, which of course depends on the ductility of the cross-sections. Since the collapse mechanism is selected at the outset using the sagging and hogging plastic moments, it is often not obvious whether the sagging hinge will form first unless an elastic analysis is undertaken. Of course, the advantage of plastic analysis is that an elastic analysis is not needed, so to safeguard against loss of ductility of the sagging hinge should this form first and require severe rotation capacity to achieve a collapse mechanism, the cross-section should satisfy Eq. 12.8.

12.3.3.2 **Example 12.3** *Analysis of a two-span continuous composite beam*

The continuous composite beam whose cross-section is shown in Figure 4.3(a) and which has the same 6 m spans with central concentrated loads that was considered in Example 12.1 will now be analysed plastically. When the beam has full shear connection, it was shown in Section 4.2.2.2(a) in Example 4.1 that $(M_p)_{sag} = 702$ kNm and in Example 12.2 that this section satisfies the sagging ductility criterion of Eq. 12.8.

Let us suppose in the hogging region (whose extent can be determined at failure from the bending moment diagram) that the concrete component has 0.6% reinforcement of yield strength $f_{yr} = 400$ N/mm² positioned 50 mm from the top, so that $P_r = 0.006 \times 130 \times 3500 \times 400$ N = 1092 kN. Noting $A_s = 9200$ mm², the rigid plastic strength of the steel component is $9200 \times 250 = 2300$ kN $> P_r$, so the neutral axis must lie in the steel component. It can be shown easily using the rigid plastic analysis techniques described in Chapter 4 and illustrated in Figure 12.6 that the plastic neutral axis lies in the top flange (so all the concrete has cracked and does not contribute to the strength) at a distance n = 145.1 mm from the top of the composite beam and that the plastic moment is $(M_p)_{hog} = 515$ kNm.

From the bending moment diagram in Figure 12.7, $W_{collapse} L/4 = (M_p)_{hog}/2 + (M_p)_{sag}$, so $W_{collapse} = 4 \times (515/2 + 702)/6 = 640$ kN. It is worth noting that the collapse load of the steel beam acting by itself was shown in Example 12.1 to be 334.5 kN. Hence the composite action has increased the strength by a factor of 640/334.5 that is by 91%. It is also worth noting that the length of the hogging region in the beam is given by A-

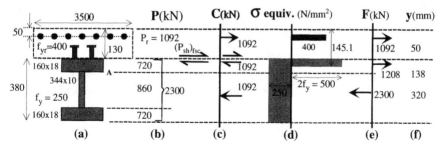

Figure 12.6 Rigid plastic analysis of composite beam in hogging region

B in Figure 12.7 as $3 \times 515/(515 + 720) = 1.25$ m and that the strength of the shear connection required in this shear span A-B is 1092 kN as shown in Figure 2.6(c). Furthermore, the strength of the shear connection required in the full sagging region span B-C in Figure 12.7 is $2 \times 2300 = 4600$ kN as can be derived in the analysis in Figure 4.4(d). Therefore, the strength of the shear connectors required in the propped cantilever of span A-C is $4600 + 1092 + 5692$ kN.

Finally, the section classification must be checked. For the compression flange outstands, $b_f = (160 - 10)/2 = 75$ and $(b_f/t_f)\sqrt{(f_y/250)} = (75/18) \times 1.0 = 4.2 < 8$ (Eq. 2.14) and so the flange is plastic. In the web from Figure 2.8 and 12.6, $y_c = 510 - 18 - 145.1 = 346.9$ and so $\alpha_c = 346.9/(344/2) = 2.0$ (Eq. 2.18). Hence using Eq. 2.19, $82/(0.4 + 0.6\alpha_c) = 82/(0.4 + 0.6 \times 2.0) = 51.3 > (d_w/t_w)\sqrt{(f_y/250)} = (344/10) \times 1.0 = 34$ and the web is also plastic. The hogging region hinge is therefore free to form and rotate. Note that the lateral-distortional buckling capacity (Chapter 13) has not been checked.

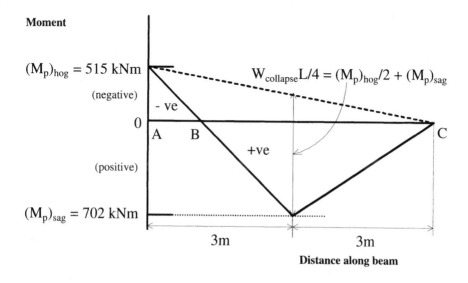

Figure 12.7 Collapse bending moment diagram for Example 12.3

12.4 References

1. Trahair, N.S. and Bradford, M.A. (1998). The Behaviour and Design of Steel Structures to AS4100. 3rd edn., E&FN Spon, London.
2. Hall, A.S., Archer, F.E. and Gilbert, R.I. (1999). Engineering Statics. UNSW Press, Sydney.
3. Warner, R.F., Rangan, B.V., Hall, A.S. and Faulkes, K. (1998). Concrete Structures. Longman, Melbourne.
4. Rotter, J.M. and Ansourian, P. (1979). 'Cross-section behaviour and ductility in composite beams'. Proceedings, Institution of Civil Engineers, London, Part 2, Vol. 67, 453-474.
5. Oehlers, D.J. and Bradford, M.A. (1995). Composite Steel and Concrete Structural Members: Fundamental Behaviour. Pergamon, Oxford.
6. Ansourian, P. (1982). 'Experiments on continuous composite beams'. Proceedings, Institution of Civil Engineers, London, Part 2, Vol. 73, 25-51.

13 Lateral-distortional buckling

13.1 Introduction

The concept of lateral-distortional buckling was introduced in Section 1.6.2. As has been noted earlier, buckling arises when portions of the steel component are subjected to compression. Lateral-distortional buckling is therefore associated with negative bending regions in composite beams, as would occur over an internal support in a continuous beam or in the region of a rigid beam to column connection. In steel structures, it is necessary to design against lateral-torsional buckling[1], and preventing this mode of failure occupies significant portions of national structural steel standards. There have been literally thousands of studies made of lateral-torsional buckling in steel beams, and it is widely accepted that the phenomenon can be predicted quite accurately. Although related to lateral-torsional buckling, the lateral-distortional buckling that occurs in composite beams is much more difficult to predict, and recourse needs to be made to advanced computer software to model it. In this chapter we will consider the concept of lateral-distortional buckling of composite beams, and consider two design approaches with the aid of examples to illustrate prediction of the buckling strength.

Lateral-distortional and lateral-torsional buckling take place when a steel section is loaded in its stronger plane, and a point is reached when the steel moves to a more favourable equilibrium position by deflecting sideways (or laterally) and twisting. The region of the beam over which this buckling takes place is usually quite long. Lateral-torsional buckling in hogging bending is shown in Figure 13.1(a), and the underlying assumption is that the cross-section remains rigid and does not distort during buckling. On the other hand, lateral-distortional buckling, as shown in Fig 1.19(a) and again in Figure 13.1(b) must be accompanied by distortion of the cross-section, since in negative bending the concrete component (although cracked) restrains the top tensile region of the steel component, and the bottom flange may only displace laterally and twist when the web element distorts in the plane of its cross-section. This distortional buckling is difficult to analyse, and the reader is directed to Ref. 2 for a review of

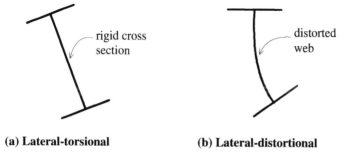

(a) Lateral-torsional   **(b) Lateral-distortional**

Figure 13.1 Lateral-torsional and lateral-distortional buckling modes

its experimental and theoretical research, and the various approaches that have been adopted.

13.2 Steel component behaviour

13.2.1 *General*

In familiar limit states terminology, lateral-distortional buckling represents an ultimate limit state, and when it occurs in a composite beam it is accompanied by catastrophic failure or so-called strain-weakening. In elastic design as in Chapter 3, it is therefore important to ascertain that the maximum negative moment in a beam is less than the strength of the composite beam as determined by lateral-distortional buckling. Generally speaking, the lateral-distortional buckling strength is not directly related to the classification of the cross-section based on local buckling that was considered in Section 2.3.3, except that it is used in its prediction of the strength of the cross-section, which is needed to determine the distortional buckling strength.

This book has considered rigid plastic analyses of cross-sections, which were introduced initially in Chapter 4. It was shown that a necessary requirement for rigid plastic analysis was that every cross-section is able to achieve the full plastic moment. Hence the steel component must not buckle locally, and this is controlled by proportioning the steel component to be at least compact, as in Section 2.3.3. This analysis philosophy must also be fulfilled by ensuring that premature lateral-distortional buckling does not occur in negative moment regions, so that the beam must be analysed for lateral-distortional buckling, and the buckling strength must not be less than the full plastic moment if a rigid plastic analysis is to be valid. In Chapter 12 we saw that rigid plastic analysis of continuous composite beams relies on moment redistribution, and it was noted that in such an analysis that lateral-distortional buckling must be prevented if this moment redistribution is to be achieved.

13.2.2   Design by buckling analysis

Design philosophy and complexity

Designing against lateral-distortional instability is usually based on procedures for steel structures, in that an elastic analysis is carried out to determine the actions in the steel portion or steel component of a cross-section subjected to negative bending. One of the complications of this analysis is that the steel component is subjected to combined negative moments and compression; the latter being in equilibrium with the tensile force in the reinforcement. The buckling strength of the steel component is then calculated, and this is used in determining the strength of the composite cross-section.

The composite section shown in Figure 13.2(a) is subjected to a negative or hogging moment. In order to demonstrate the complexity of this buckling problem, let us consider the results of a rigid plastic analysis, which is described in detail in Section 12.3.3, and which are depicted in Figures. 13.2(b) and (c). It can be seen that the resultant compressive force in the steel component F_{comp} is equal to the tensile strength of the reinforcing bars $F_{ten} = P_r = A_r f_{yr}$. The force F_{comp} is resisted by the compressive stresses adjacent to the centroid of the steel component as shown in (c) which leaves the stresses in the rest of the steel component to resist the moment in the steel component M_{steel} as shown. It can therefore be seen in (b) that the steel component is subjected to a moment and axial force both of which vary along the length of the beam and both of which vary according to the variation of the degree of shear connection η along the length of the beam, as described in Chapter 4. These variations make the buckling problem in composite beams much more complicated than in a steel beam.

13.2.2.2 Elastic buckling parameters M_{od} and N_{od}

The first step in a thorough elastic analysis of lateral distortional buckling is to determine the moment M_{steel} in Figure 13.2(b) which will cause buckling when the compressive force F_{comp} is not present. This moment capacity is referred to as the elastic lateral-distortional buckling moment in the steel section M_{od}, and its derivation is computationally difficult. Fortunately, there are design methods that can be used either directly for this calculation, or which make use of the lateral-distortional buckling moment M_{od} implicitly, and these will be treated in Section 13.3.

As was noted in Section 13.2.2.1 and in Figure 13.2(b), the steel component is also subjected to an axial compressive force F_{comp} that equilibrates the tension in the reinforcement. The force F_{comp} to cause buckling in the absence of M_{steel} in (b) also needs to be determined in an accurate distortional buckling analysis. This force is referred to as the elastic lateral-distortional buckling compressive load N_{od}. The simplified design method of Section 13.3.2 allows this load to be calculated, but rigorous incorporation of the effects of the axial force is often omitted as the stresses induced in the bottom compressive flange due to bending are usually much larger than those induced by the axial compression[3], as will be shown subsequently.

13.2.2.3 *Strengths in bending and compression M_{sd} and N_{sd}*

In the method of 'design by buckling analysis'[1], the elastic buckling moment M_{od} and the elastic buckling load N_{od} for the steel component are converted into *strengths* using relevant strength curves in national standards. Typical illustrations of these

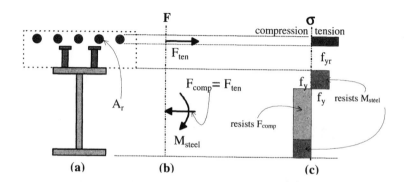

Figure 13.2 Composite beam subjected to a negative or hogging moment

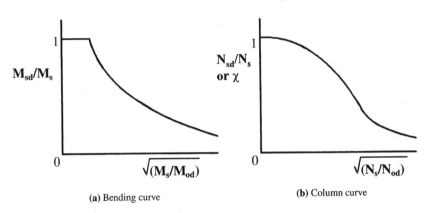

(a) Bending curve

(b) Column curve

Figure 13.3 Buckling curves in bending and compression

curves are shown in Figures. 13.3(a) and (b), where the non-dimensional slenderness is written as $\sqrt{(M_s/M_{od})}$ for bending and $\sqrt{(N_s/N_{od})}$ for compression, where M_s is the cross-section strength of the steel component in bending (which is its plastic moment M_{ps} if the cross-section is compact) and N_s is the cross-section strength of the steel component in compression (which is its squash load N_{sq} if the cross-section is free from local buckling effects). It is worth noting that these curves resemble the complex interaction between buckling and non-linear material behaviour in steel columns as shown in Figure 8.1. Hence as in the failure envelope for the steel columns in Figure 8.1, these curves relate the pure axial strength and pure flexural strength with both the elastic and rigid plastic strengths.

Once these slendernesses have been determined using the elastic distortional buckling moment M_{od} and load N_{od}, the bending strength of the steel in the absence of compression M_{sd} and the compressive strength of the steel in the absence of bending N_{sd} may be determined from national standards. For example, we saw in Section 8.2.1.3 that the strength of the steel component in compression in accordance with the Eurocode was

$$N_{sd} = \chi N_s \tag{13.1}$$

for an appropriate column curve in which χ is a function of N_{od}, while in the Australian AS4100 steel standard the bending strength of the steel component is written as

$$M_{sd} = 0.6\left\{\sqrt{\left(\frac{M_s}{M_{od}}\right)^2 + 3} \ - \ \left(\frac{M_s}{M_{od}}\right)\right\}M_s \ \leq M_s \tag{13.2}$$

It must be noted at this stage that Eqs. 13.1 and 13.2 were developed for lateral-torsional beam buckling and flexural column buckling. Their use for lateral-distortional buckling is questionable, but they are considered to be conservative and their use will therefore be recommended.

13.2.2.4 *Interaction between axial load and flexure M_{sdr}*

Finally, the bending strength must be reduced for the effects of axial compression if an 'accurate' analysis is being performed by treating the steel component as a beam-column. Because we are determining the buckling moment of a steel cross-section, the compression in the steel component is not known at the outset. Although an elastic analysis is being undertaken for the buckling analysis, it is conservative to use the yield strength of the reinforcement $f_{yr}A_r$ in the concrete component as the compression in the steel component. Hence and in accordance with national steel standards, the steel member strength M_{sdr} can be determined by reducing M_{sd} according to

$$M_{sdr} = M_{sd}\left(1 - \frac{A_r f_{yr}}{N_{sd}}\right) \tag{13.3}$$

It is worth reiterating firstly that the calculation of the lateral-distortional buckling moment and load are based on very approximate design models in lieu of complex finite element modelling, and secondly that the member strengths that are based on combined elastic buckling and yielding are derived from lateral buckling results, and their applicability to lateral-distortional member strengths is questionable. Because of this, design may often be simplified by ignoring the effects of compression. In Sect. 13.3.2, the U-Frame Method will be used to show in principle how compression may be incorporated into the buckling analysis, while in the Alternative Method in Sect. 13.3.3 it will be ignored.

13.3 Design models

13.3.1 *General*

There are a number of ways of treating lateral-distortional buckling in composite beams, and advanced finite element software that can in principle handle the phenomenon is now becoming available. However, this software is only a research tool, and it is easier and more appropriate to use design equations. The so-called Inverted U-frame Method and an alternative method based on finite element studies will be considered in this section.

13.3.2 *Inverted u-frame approach*

13.3.2.1 Elastic buckling

The Inverted U-frame Approach is based on the design philosophy for half-through girder bridges. Consider the inverted U-frame shown in Figure 13.4. The compression flange is modelled as a uniformly compressed strut restrained elastically against flexural buckling by the stiffness of the web. The web is treated as a cantilever, and its stiffness may be determined by applying the fictitious unit horizontal forces shown. In addition, the unit horizontal forces may cause bending of the composite beam between the webs of the beams, so a rotational stiffness at the top or tension flange/ web junction of $2(EI)_{slab}/L_b$, in which $(EI)_{slab}$ is the flexural rigidity of the slab and L_b is the length of the slab between the webs of the parallel beams as shown, may be included. However, this effect is fairly small and may be ignored.

Figure 13.5 shows the strut buckling model, in which the flange strut is subjected to an elastic restraint of stiffness α_t per unit length that produces a distributed restoring force of $\alpha_t u_t$ per unit length, where u_t is the buckling deformation which is assumed to be a sine curve. It can be shown readily that the elastic critical value of the force in the strut to cause buckling N_{cr} is

$$N_{cr} = \frac{\pi^2 E_s I_F}{L^2} + \frac{\alpha_t L^2}{\pi^2} \tag{13.4}$$

where I_F is the second moment of area of the flange about the web. The relationship between N_{cr} and L in Eq. 13.4 is of a garland shape, and the minimum value of N_{cr} may be determined by setting dN_{cr}/dL to zero. Hence,

$$\left(N_{cr}\right)_{min} = 2\sqrt{E_s I_F \alpha_t} \tag{13.5}$$

The conversion of $(N_{cr})_{min}$ to determine M_{od} for the steel component and then to a composite member strength will be illustrated in the following example.

13.3.2.2 **Example 13.1** *Beam strength using the Inverted U-Frame Approach*

(a) $F_{comp} = A_r f_{yr}$

Consider the beam analysed in Example 3.3 and shown in Figure 3.2 when subjected to negative bending. The slab in this example had 0.6% reinforcement ($A_r = 1170$

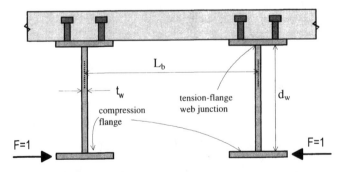

Figure 13.4 Inverted U-frame

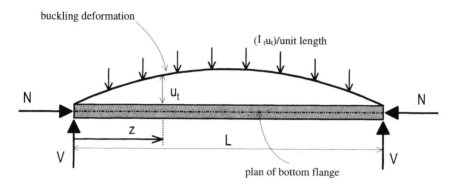

Figure 13.5 Strut model

mm²). We will assume that the web acting as a cantilever has a length of 300 + 12/2 = 306 mm. Applying a unit force of 1 N/mm length to the centroid of the bottom flange produces a deformation in the web acting as a cantilever (whose second moment of area is $7^3/12 = 28.58$ mm⁴/mm length) of $306^3/(3×200×10^3×28.58) = 1.67$ mm. The stiffness α_t is the force (1 N/mm) divided by the deformation (1.67 mm), so that $\alpha_t = 0.598$ N/mm². The second moment of area of the bottom flange is $I_f = 170^3×12/12 = 4.913×10^6$ mm⁴, and so substituting into Eq. 13.5 produces $(N_{cr})_{min} = 2\sqrt{(200×10^3×4.913×10^6×0.598)}$ N = 1533 kN.

The *elastic* critical load of $(N_{cr})_{min} = 1533$ kN in the compression flange must now be converted to the buckling *strength* of the steel component M_{sd} by firstly calculating the elastic distortional buckling moment M_{od}. Firstly, this load produces a stress in the compression flange of $(N_{cr})_{min}/A_f = (1533×10^3)/(170×12) = 752$ N/mm², and using from Figure 13.6 the value of $I_s = 115.1×10^6$ mm⁴ produces (from simple linear elastic beam theory $M = \sigma I/y$) an elastic buckling moment of $M_{od} = (752×115.1×10^6)/(150+6)$ Nmm = 554.8 kNm. Secondly, we require the cross-section strength M_s for the steel component. It can be shown (although the position of the neutral axis for the *composite* section is needed to determine the web parameter α_c in Eq. 2.19) that the steel component is compact, and has a plastic section modulus about its major axis, using the procedure illustrated in Figure 4.5, of $S_s = 794×10^3$ mm³. For the yield stress $f_y = 300$ N/mm² (as assumed in Example 3.4), $M_s = f_y S_s = 300×794×10^3$ Nmm = 238.2 kNm and substituting into Eq. 13.2 produces $M_{sd} = 0.6 × (\sqrt{((238.2/554.8)^2 + 3)} - (238.2/554.8)) × 238.2 = 193.7$ kNm.

Finally, we will assume that the reinforcement (positioned 35 mm below the top surface as in Example 3.3 and Figure 13.6) is at yield at $f_{yr} = 400$ N/mm² which results in a tensile force of $F_{ten} = A_r f_{yr} = 1170 × 400$ N = 468 kN. The stress resultants across the composite section have the distribution shown in Figure 13.6(b) which is the same as that shown in Figure 1.16(d) except that the signs are reversed because we are dealing with a negative region instead of the positive region depicted in Figure 1.16. Hence taking moments about the centroid of the steel component in Figure 13.6(b) produces the moment capacity of the composite section when governed by distortional buckling

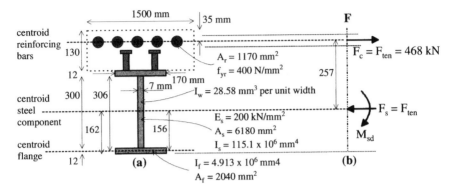

Figure 13.6 Composite beam subjected to lateral-distortional buckling

as M_{cmp} = 193.7 + (130 − 35+12+300/2) × 468/1000 = 314.0 kNm. Note that this buckling moment is greater than the first yield moment of 266.2 kNm determined in Example 3.4 for the composite beam in hogging bending, so that first yield is attainable prior to lateral-distortional buckling. We have also assumed full shear connection.

(b) F_{comp} from linear elastic analysis

Using the more accurate linear elastic method of Example 3.3 and 3.4, at buckling of the steel component the stress in the bottom fibre is $\sigma_b = M_{sd}y/I_s = 193.7 \times 10^6$ × (150 + 12)/(115.1 × 10⁶) = 272.6 N/mm² for which the curvature (from Example 3.4) is $\kappa = \sigma_b/E_sn_b$ = 272.6/(200 × 10³ × 203) = 6.71 × 10⁻⁶ mm⁻¹ and which results in a moment in the composite beam of $M_{steel} = E_sI_{ns}\kappa$ = 200 × 10³ × 180.1 × 10⁶ × 6.71 × 10⁻⁶ Nmm = 241.7 kNm. The 30% disparity between the moment of 241.7 kNm that was based on the linear elastic analysis of Examples 3.3 and 3.4 and the moment of 314.0 kNm that was determined assuming that the reinforcement was at yield illustrates the unconservatism of the latter assumption, although the calculation is easier than the linear elastic method. It is worth noting that at the curvature of 6.71 × 10⁶ mm⁻¹ at which the beam buckles in a lateral-distortional mode, the stress in the reinforcement is 6.71 × 10⁻⁶ × (454 − 35 − 203) × 200 × 10³ = 291.2 N/mm² which is significantly less than the yield value of 400 N/mm² assumed.

It is also worth noting from the linear elastic analysis and the lateral-distortional buckling analysis of the steel component, that the stress in the bottom flange at buckling is 272.6 N/mm². The tensile stress in the reinforcement at buckling (initially assumed to be zero in the buckling analysis) also produces a compressive stress in the bottom flange of 291.2 ×A_r/A_s = 291.2 × 1170/6180 = 55.1 N/mm². This additional axial stress is 20% of the bending stress at buckling based on pure bending of the steel component, and illustrates that an accurate analysis must treat the steel component as a beam-column. Although recourse could be made to Eq. 13.3 to allow for the additional compression in the steel component that equilibrates with the tension in the reinforcement, the highly conservative assumptions of the Inverted U-Frame Approach probably do not justify the additional effort.

13.3.3 *Empirical approach*

13.3.3.1 Inelastic finite element solutions

The use of Eq. 13.2 showed that the governing variable for determining the buckling strength of the steel component in hogging bending was $\lambda_d = \sqrt{(M_s/M_{od})}$, and we have demonstrated already the difficulty in calculating M_{od} accurately. Of course, the section strength M_s of the steel component is readily obtainable. A sophisticated finite element analysis that incorporated geometric and material nonlinearities was undertaken to determine the strength of the steel component of a composite beam when governed by lateral-distortional buckling by Weston, Nethercot and Crisfield[4]. Their study was restrictive, owing to the immense computational prowess that was needed to obtain numerical solutions. Bradford[5] also analysed the inelastic lateral-distortional buckling of steel beams restrained by a slab, using a computer program that required less computing effort than the study of Weston *et al.* Bradford's solutions compared favourably with those of Weston *et al.* when their modified slenderness λ_d was adjusted slightly to

$$\lambda_d = 0.018\left(\frac{L_d}{r_y}\right)^{\tfrac{1}{2}}\left(\frac{d_w}{t_w}\right)^{\tfrac{1}{3}} - 0.40 \qquad (13.6)$$

and the strength was predicted by modifying Eq. 13.2 to

$$M_{sd} = 0.8\left\{\sqrt{\lambda_d^4 + 3} - \lambda_d^2\right\}M_s \;\; \leq \;\; M_s \qquad (13.7)$$

in which L_d is the length of the beam between supports, r_y is the minor axis radius of gyration of the compression flange, and d_w and t_w are the depth and thickness of the web respectively. The solutions were obtained with the yield stress f_y in the range 250 N/mm² to 400 N/mm² for which the prediction of the modified slenderness in Eq. 13.6 was found to be virtually independent of M_s and hence f_y. The use of Eqs. 13.6 and 13.7 should therefore be limited to steel elements whose yield stresses are in the range indicated, and implicit in the statement of Eq. 13.6 is the value of the elastic distortional buckling of the steel component M_{od}.

13.3.3.2 **Example 13.2** *Beam strength using the Empirical Approach*

The beam analysed in Example 13.1 will now be analysed using Eqs. 13.6 and 13.7. It was shown in the Inverted U-Frame method that the buckling moment was independent of the beam length, so here we will assume $L_d = 7$ m. Thus $I_f = 4.913 \times 10^6$ mm⁴, $A_f = 170 \times 12 = 2040$ mm², $r_y = \sqrt{(4.913 \times 10^6/2040)} = 49.1$ mm, and $\lambda_d = 0.018 \times (7000/49.1)^{1/2}(300/7)^{1/3} - 0.40 = 0.352$. Hence from Eq. 13.7 and using $M_s = 238.2$ kNm (from Example 13.1), $M_{sd} = 0.8 \times ((\sqrt{(0.352^4 + 3)} - 0.352^2) \times 238.2 = 307.3 > M_s$, and so $M_{sd} = 238.2$ kNm and the steel component and hence the composite beam is unaffected by lateral-distortional buckling.

Finally, we may make the simple assumption, albeit unconservative, that the reinforcement has reached yield at 468 kN, so that $M_{cmp} = 238.2 + (130 - 35 + 12 + 300/2) \times 468/1000 = 358.5$ kNm. The buckling capacity of the steel component using the Empirical Method is thus 238.2/193.7 or 23% greater than that of the Inverted U-Frame Method, and the corresponding composite beam bending strength is 358.5/314.0 or 14% greater. The comparisons are somewhat arbitrary, however, as the Empirical Approach depends on the beam length and the Inverted U-Frame Approach does not. It is worth noting that in this example the latter model is more conservative than the former, and this is generally the case for practical beams.

13.4 Recommendations

The discussion and examples presented in this chapter have illustrated the substantial difficulties that arise when modelling lateral-distortional buckling in composite beams. Nevertheless, a necessary requirement is that this mode of buckling must be prevented if a rigid plastic analysis of continuous composite beams is to be carried out.

Although there are a number of approaches to the problem, it is recommended that the Inverted U-Frame Approach illustrated in Example 13.1 be used to determine the bending capacity of the steel component through the use of Eq. 13.2 by firstly converting N_{cr} in Eq. 13.4 to the elastic distortional moment M_{od}. The neutral axis can be determined by the elastic methods of Chapter 3 and the curvature at buckling then calculated from the strain in the bottom compressive flange of the steel section at buckling. The capacity of the composite beam is then simply the product of its elastically-determined flexural rigidity and the curvature, which is conservative.

The recommended method described above does not allow for the destabilizing compressive force in the steel member that equilibrates with the tension in the reinforcement. In theory this effect, which lowers the buckling moment, can be included by treating the steel component as a beam-column with respect to buckling. However, this is not considered necessary as any lack of conservatism produced by ignoring the compressive force is more than compensated for by the assumption in the U-Frame Model that the flange is a uniformly compressed strut, when in fact the compression in the flange-strut idealization varies in accordance with the bending moment distribution along the composite beam.

13.5 References

1. Trahair, N.S. and Bradford, M.A. (1998). The Behaviour and Design of Steel Structures to AS4100. 3rd edn., E&FN Spon, London.
2. Bradford, M.A. (1992). 'Lateral-distortional buckling of steel I-section members'. Journal of Constructional Steel Research, Vol. 23, 97-116.
3. Bradford, M.A. and Ronagh, H.R. (1997). 'Generalised elastic buckling of restrained I-beams by the FEM'. Journal of Structural Engineering, ASCE, Vol. 123, No. 12, 1361-1367.
4. Weston, G., Nethercot, D.A. and Crisfield, M. (1991). 'Lateral buckling in continuous composite bridge girders'. The Structural Engineer, Vol. 69, No. 5, 79-87.
5. Bradford, M.A. (1989). 'Buckling strength of partially restrained I-beams'. Journal of Structural Engineering, ASCE, Vol. 115, No. 5, 1272-1276.

14 General fatigue analysis procedures

14.1 Introduction

Composite beams in buildings and particularly in bridges are often required to resist continuous repetitions of applied loads such as those imposed by the traversal of vehicles or cranes. Although these fatigue loads are small in comparison with the ultimate strength of the structure, the *tensile* component of the stresses induced by these fatigue loads can cause cracks to initiate and propagate that can eventually lead to failure of the structure at serviceability loads. It is, therefore, necessary for the engineer to ensure that the remaining strength or residual strength of the structure during the whole design life of the structure is greater than the maximum possible design overload.

In this chapter, generic fatigue analysis procedures and behaviours are described that can be applied to any type of component. It is assumed throughout that the composite beam behaves in a full interaction linear elastic fashion. General forms of the fatigue material properties are first described, followed by methods for quantifying the numerous applications of load and the cyclic stress resultants that they induce, and finally a generic form of the fundamental fatigue equation is developed. In Chapter 15, the generic fatigue equation is applied specifically to the assessment and design of stud shear connectors in composite beams.

14.2 General fatigue properties

14.2.1 General

There are two fundamental properties that are required for fatigue analysis which are the *endurance* and the *residual strength* of the structural component. The endurance is a measure of the rate of fatigue damage, such as crack initiation or crack propagation, whereas, the residual strength is a measure of the effect of the fatigue damage on the remaining static strength of the structure.

14.2.2 Fatigue endurances

The endurance of a structural component E, that is the number of cycles to failure, depends on the range R of the cyclic load that induces tensile stresses in the component. The dependence of the endurance on the range of the cyclic load is usually determined experimentally and the relationship derived from a linear regression of the logarithm of the variables as shown in Figure 14.1. To allow for the scatter of the test results, design is usually based on the characteristic endurance at 2 standard deviations, that is at a 2.3% probability of failure, which is given by the following equation

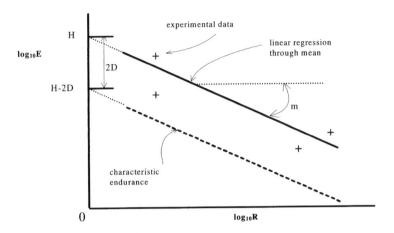

Figure 14.1 Fatigue endurance

$$E_{ch} = 10^{H-2D} R^{-m} \tag{14.1}$$

where m is the slope of the regression line in Figure 14.1, H the intercept with the y-axis and D is the magnitude of one standard deviation.

14.2.2.1 **Example 14.1** *General fatigue endurance equation*
(a) Material component parameter m
It can be seen in Eq. 14.1 that the endurance is inversely proportional to R^m. The exponent m is a material property that varies from 3 for welded steel components to about 20 for concrete, with stud shear connectors having a value of about 5. The exponent m defines the susceptibility of the component to changes in the cyclic range R. For example, halving the range of the cyclic load that is applied to a welded steel component will increase the endurance by a factor of $2^3 = 8$. Whereas, halving the cyclic range that is applied to a concrete component will increase the endurance by a factor of $2^{20} \approx 1,000,000$. Hence, a concrete component is much more sensitive to changes in the applied range than welded steel components as it has a larger value of the exponent m.

(b) Component detail parameter H
It can also be seen in Eq. 14.1 that the mean endurance is directly proportional to 10^H where the intercept-constant H in Figure 14.1 defines the susceptibility of the structural component detail to fatigue failure. For example for a steel flange plate with a smoothly varying cross-sectional shape, $10^H \approx 2 \times 10^{15}$ when the stress is measured in N/mm^2. However, when a small hole is inserted into this flange, then the stress concentrations caused by the hole reduce the constant to $10^H \approx 4 \times 10^{12}$, that is the endurance has reduced by a factor of $(2 \times 10^{15})/(4 \times 10^{12}) = 500$. Furthermore, welding a stud shear

connector to the flange will not only induce stress concentrations but also cause minute cracks and residual stresses that reduce the constant to $10^H \approx 2 \times 10^{12}$ and, hence, reduce the fatigue life of the original flange by a factor of about 1000.

(c) Standard deviation D

It is also worth noting that the scatter associated with predicting fatigue endurances is very large. For example for steel components, the characteristic endurance as a proportion of the mean endurance that is $10^{H-2D}/10^H \approx 0.4$. This means that a component that is designed for a mean fatigue life of 100 years has a 4.6% probability of failing at less than 40 years or more than 250 years.

14.2.3　　*Residual strength*

14.2.3.1 Accumulated damage laws

Two forms of residual strength envelopes are shown in Figure 14.2, where P_s is the static strength of a component prior to cyclic loading, P_c is the remaining or residual strength after cyclic loading, N_k is the number of cycles of load of range R_k that have been applied, and E_k is the endurance of the component at the range R_k. The residual strength variation marked 'B' will be referred to as the 'crack initiation approach'; this variation assumes that there is no reduction in strength until $N_k > E_k$ and is applicable to steel components, particularly when a large number of cyclic loads are required to initiate a crack. In contrast, the residual strength variation marked 'A' assumes a linear reduction in the strength as soon as cyclic loads are applied and will be referred to as the 'crack propagation approach'. This approach is applicable to stud shear connectors where tests have shown that minute cracks in the weld zones are propagated immediately cyclic loads are applied and cause an immediate reduction in strength.

(a) Crack propagation approach

Let us first consider the crack propagation linear residual strength envelope 'A' in Figure 14.2. When the number of applied cycles of load $N_k = 0$, the residual strength P_c is equal to the static strength P_s, that is, the strength prior to cyclic loading. Furthermore when $N_k = E_k$, then the residual strength $P_c = 0$. The residual strength envelope is, therefore, given by $N_k/E_k = 1 - P_c/P_s$. It can be shown[1] that because the residual strength envelope is linear, it can be applied to combinations of ranges of cyclic loads and, furthermore, that the sequence in which the cyclic loads are applied does not affect the reduction in strength. In which case, the linear residual strength failure envelope is given by the following equation with an inequality.

$$\sum_{k=1}^{k=z} \frac{N_k}{E_k} \leq 1 - \frac{P_c}{P_s} \tag{14.2}$$

where there are a total of z magnitudes of the ranges of applied loads.

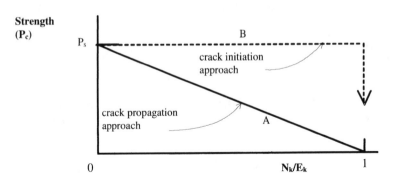

Figure 14.2 Residual strength

(b) Crack initiation approach

The crack initiation approach failure envelope 'B' in Figure 14.2 is a special case[1] of Eq. 14.2 because when

$$\sum_{k=1}^{k=z} \frac{N_k}{E_k} \leq 1 \tag{14.3}$$

then $P_c = P_s$.

(c) Fatigue damage

Equations 14.2 and 14.3 are also referred to as accumulated damage laws. The parameter N_k/E_k on the left hand side of these equations can be considered to be the fatigue damage that the cyclic loads have induced. Whereas, the right hand side of these equations is the fatigue damage that can be sustained. Once the left hand side exceeds the right hand side then the structure fails as the residual strength can no longer resist the applied load. It is also worth noting that the fatigue damage term N_k/E_k in Eq. 14.2 is proportional to $N_k R^m$, as $E \propto R^{-m}$ in Eq. 14.1. Hence, the fatigue damage is directly proportional to both the number of cycles applied and the magnitude of the range of the cyclic load.

14.2.3.2 *Fatigue analysis procedures*

The accumulated damage laws of Eqs. 14.2 and 14.3 are fundamentally different, and this difference affects their application in the design or assessment of composite bridge beams. These fundamental analytical differences are demonstrated by considering the fatigue design of the shear connection in a simply supported composite beam in Figure 14.3(a) that is subjected to the longitudinal traversal of a point load.

The longitudinal traversal of a point load across a simply supported composite

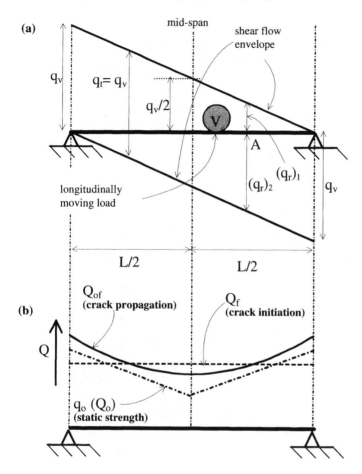

(a)

mid-span

shear flow envelope

q_v

$q_t = q_v$

$q_v/2$

V

A

longitudinally moving load

$(q_r)_1$

$(q_r)_2$

q_v

L/2

L/2

(b)

Q_{of} (crack propagation)

Q_f (crack initiation)

Q

$q_o (Q_o)$ (static strength)

Figure 14.3 Fatigue analysis procedures

bridge beam produces the shear flow envelope shown in Figure 14.3(a). At a design point such as at 'A', the shear connectors are subjected to uni-directional shear flow forces of $(q_r)_1$ and $(q_r)_2$ and, hence, a total cyclic range of $q_t = (q_r)_1 + (q_r)_2$. For this example of a point load moving across a simply supported beam, the total range of the shear flow force q_t is constant throughout the span of the beam and equal to the maximum uni-directional shear flow force at the supports q_v. Furthermore, the uni-directional shear flow force at mid-span is equal to half of the uni-directional shear flow force at the supports, that is $q_v/2$. As we are dealing with stud shear connections, it will be assumed that the endurance is a function of the total range q_t as will be shown in Section 15.2.

(a) Overload

Irrespective of whether the crack initiation or crack propagation procedure is being applied, the engineer has to ensure that the maximum static design load or maximum overload can be resisted. Therefore, shear connectors have to be distributed along the beam in proportion to the variation along the beam of the maximum uni-directional shear flow force. This is shown in Figure 14.3(b) as the shear flow resistance Q_o that is required to resist the maximum overload. The static strength Q_o is proportional to q_v at the supports and proportional to $q_v/2$ at mid-span.

(b) Crack initiation procedure

The crack initiation approach, as given by the accumulated damage law of Eq. 14.3 and illustrated in Figure 14.2 as line B, assumes that there is no reduction in strength until the fatigue endurance of the component is exceeded when $N_k > E_k$. It is, therefore, necessary to ensure that the fatigue endurance is never exceeded, that is $N_k < E_k$ at any position of the beam. It will be shown in Section 14.6 how Eqs. 14.1 and 14.3 can be developed to determine the shear flow strengths of the shear connectors Q_f in order to ensure that the fatigue cyclic loads do not cause failure of the shear connectors during the design life. As the total range q_t causes fatigue damage in shear connectors and as this is constant over the length of the beam as shown in Figure 14.3(a), then the shear flow strength required to ensure fatigue failure does not occur is also constant along the length of the beam as shown by Q_f in Figure 14.3(b).

The engineer still has to ensure that the maximum static load can be resisted, so that the shear flow strength of the shear connectors must be equal to or exceed Q_o in Figure 14.3(b). Therefore, the upper bound of Q_o and Q_f is used in the crack initiation design procedure. It can be seen in Figure 14.3(b) that for this analysis, the static strength requirement governs the design near the supports, whereas, the fatigue requirement governs the design near mid-span.

(c) Crack propagation procedure

The crack propagation procedure that is given by the accumulated damage law of Eq. 14.2 and illustrated by line A in Figure 14.2, assumes that the residual strength is dependent on the fatigue loads. In this approach, the static strength requirement Q_o in Figure 14.3(b) is increased to Q_{of} to allow for the anticipated reduction in strength due to the fatigue damage during the design life of the structure. Hence, the shear flow strength Q_{of} is the strength required when the structure is first built and Q_o is the anticipated strength at the end of the design life that is sufficient to resist the maximum overload. The difference $Q_{of} - Q_o$ is the anticipated reduction in strength due to fatigue damage during the design life.

In summary, the crack initiation approach requires two completely separate and independent analyses for strength and endurance that assume that fatigue loads do not affect the static strength until the fatigue endurance has been reached. In contrast, the crack propagation approach requires one analysis in which the strength and endurance are integrally related.

14.3 Applied loads on bridges

14.3.1   *General*

A composite bridge beam must be able to resist the maximum overload or ultimate strength design load that may occur infrequently, such as once in a design life time or not at all. The composite bridge beam must also be able to sustain the fatigue damage induced by all of the working or serviceability vehicular loads that can be applied to the composite beam as frequently as several times a minute. Furthermore, it is necessary to ensure that the residual strength of the composite beam, after any reduction in strength due to fatigue damage, can resist the maximum overload during the whole design life of the composite beam.

There is an enormous variety of shapes of vehicles, an enormous variety of weights of these vehicles, and an enormous variety of combinations of these weights and shapes. It would be impossible to determine the stress resultants from each individual combination. Therefore, the problem is simplified by designing for a specific number of vehicle traversals, using standard vehicular shapes (that is standard fatigue vehicles), and standard combinations of weights (that is load spectrums). The problem is further simplified by assuming the structure behaves in a linear elastic fashion which will allow the principle of superposition to be used; this is a reasonable assumption as fatigue damage is induced by working or serviceability loads. A further assumption that is often used for composite bridge beams is to assume that there is full interaction between the concrete and steel components; this assumption simplifies the determination of the stress resultants but tends to overestimate the shear flows and underestimate the flexural stresses.

14.3.2   *Frequency of fatigue vehicles*

A bridge may be subjected to vehicles that range in weight from cars of a few kN to commercial vehicles weighing 4000 kN. It was shown in Section 14.2.3.1 that the fatigue damage is proportional to $N_k R^m$. As we are assuming linear elastic behaviour, the range of the cyclic load R is directly proportional to the weight of the fatigue vehicle W_{FV}. Therefore, the fatigue damage is proportional to $N_{FV}(W_{FV})^m$ where N_{FV} is the number of traversals of a fatigue vehicle of weight W_{FV} and the exponent m can range from 3 to 6. Even though cars occur much more frequently than commercial vehicles, the fatigue damage that they induce is much less than that induced by commercial vehicles and, therefore, cars are generally ignored in a fatigue analysis which is usually restricted to the traversal of commercial vehicles.

14.3.2.1 **Example 14.2** *Load traversals*

A bridge may be subjected to 2.5 million commercial vehicle traversals per year from one lane, 2 million traversals per year in an adjacent lane, and may be required to last 120 years. Hence, a component in a beam that is loaded from both lanes will be subjected to a total number of fatigue vehicle traversals of $T = (2.5 \times 120) + (2 \times 120) = 540$ million. Furthermore, as each vehicle traversal can induce serval ranges of cyclic load, as will be shown in Section 14.4, the component may have to be designed to resist over a billion cycles of stress. It can now be seen that structural components in bridges can be subjected to an enormous number of cyclic stresses, which is the reason fatigue failure is the most common form of failure in bridges.

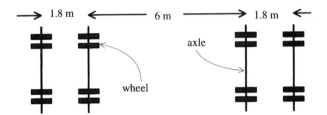

Figure 14.4 Standard fatigue vehicle

14.3.3   *Standard fatigue vehicles*

A typical example[2] of the axle distribution of a *standard fatigue vehicle* that is used to represent all the fatigue vehicles that traverse the composite beam is shown in Figure 14.4. The shape of the vehicle, that is the number of axles and their distribution, should be chosen to induce the same fatigue damage as the fatigue vehicles of the same weight. If necessary, several shapes of standard fatigue vehicles can be used to represent the myriad shapes of fatigue vehicles.

A typical weight of a standard fatigue vehicle[2] is 20 kN per wheel, which for the standard fatigue vehicle in Figure 14.4 gives a weight of $W_{SFV} = 320$ kN. The actual weight of the standard fatigue vehicle W_{SFV} that is used in the fatigue analysis is not important as variations in the weight can be allowed for in a *load spectrum* that is described in the following section. The standard fatigue vehicle can be used to traverse the bridge to determine the theoretical cyclic stress resultants as described in Section 14.4. As the bridge is assumed to be behaving in a linear elastic fashion, the traversal of the standard fatigue vehicle can be used to determine the fatigue damage for all the fatigue vehicles as described in Section 14.6.

14.3.4   *Load spectrum*

Each standard fatigue vehicle can be used to represent a group of fatigue vehicles of varying weights. It is useful practice to represent the variation in the fatigue vehicle weights W_{FV} as a proportion of the weight of the standard fatigue vehicle W_{SFV}. This is shown as W in column 2 in Table 14.1, where there are 'i' weights of fatigue vehicles as shown in column 1. The weight of a fatigue vehicle at level x is, therefore, $(W_{FV})_x = W_x W_{SFV}$. The probability of occurrence of each weight of vehicle at each of the 'i' levels in Table 14.1 is given as B in column 3, such that the summation of B is unity. Therefore, the number of fatigue traversals at level x is $B_x T$, where T is the total number of load traversals from all of the fatigue vehicles associated with the standard fatigue vehicle.

Columns 2 and 3 in Table 14.1 are often referred to as a load spectrum. The fatigue damage has already been shown in Section 14.3.2 to be proportional to $N_{FV}(W_{FV})^m$. As the number of vehicle traversals at level x is $N_x = B_x T$ and the weight of the vehicle $(W_{FV})_x \propto W_x$, then the fatigue damage is proportional to $B_x W_x^m$ which is given in column 4 in Table 14.1.

14.3.4.1 **Example 14.3** *Frequency of fatigue loads*

Let us assume that: the weight of the standard fatigue vehicle is $W_{SFV} = 320$ kN; the load spectrum is given by columns 2 and 3 in Table 14.2, which is similar to
Table 14.1 Format of load spectrum

Level (x)	Weight (W)	Probability (B)	BW^m
(1)	(2)	(3)	(4)
1	W_1	B_1	$B_1W_1{}^m$
2	W_2	B_2	$B_2W_2{}^m$
i	W_i	B_i	$B_iW_i{}^m$
		$\sum = 1^\varepsilon = 1$	$L_f = \sum BW^m$

that applied to a British motorway[2]; the design life of the bridge is 120 years; and that the structure in which the component is located is subjected to T = 240 million applications of fatigue vehicle traversals during its design life.

From Table 14.2, the weights of the fatigue vehicles range from $6.5 \times 320 = 2080$ kN at level 1 to $0.2 \times 320 = 64$ kN at level 6. The number of load traversals at level 1 for the vehicle of weight 2080 kN is $B_1T = 4800$ which is just over three per month. In contrast, the number of load traversals for the lightest fatigue vehicle of weight 64 kN at level 6 is 144 million which is just over two per minute.

Table 14.2 Load spectrum

Level	W	B	BW^3 welded components		BW^4 non-welded components		$BW^{5.1}$ stud shear connectors		BW^{20} concrete	
(1)	(2)	(3)	(4)	%	(5)	%	(6)	%	(7)	%
1	6.5	0.00002	0.006	2	0.036	9	0.280	25	363×10^9	97
2	5.0	0.00010	0.013	5	0.063	15	0.367	32	10×10^9	3
3	2.0	0.01000	0.080	29	0.160	39	0.343	30	0	0
4	1.0	0.13988	0.140	51	0.140	34	0.140	12	0	0
5	0.5	0.25000	0.031	11	0.016	4	0.007	1	0	0
6	0.2	0.60000	0.005	2	0.001	0	0.000	0	0	0

$$W_{SFV} = 320\text{kN} \qquad \Sigma = 0.274 = L_f \qquad \Sigma = 0.415 = L_f \qquad \Sigma = 1.14 = L_f$$

14.3.4.2 **Example 14.4** *Distribution of fatigue damage in a load spectrum*

It was shown in Section 14.3.4 that the fatigue damage is proportional to $B_x W_x{}^m$. For the load spectrum of columns 2 and 3 in Table 14.2, the fatigue damage at each level of the load spectrum is shown in column 4 for the case of a welded component where m = 3. The sum of the fatigue damage terms in column 4 is 0.274 as shown in the bottom row. The fatigue damage terms in column 4 are also given as a proportion of the total fatigue damage of 0.274 in the adjacent column. Similar analyses have been applied to a non-welded component in column 5 where m = 4, stud shear connectors in column 6 where m = 5.1, and to a concrete component in column 7 where it is assumed that m = 20.

By comparing columns 4 to 7 in Table 14.2, it can be seen that as the exponent m increases the greatest fatigue damage occurs at greater fatigue vehicle weights. For example for the welded component in column 4 in which m = 3, 51% of the fatigue damage occurs for vehicles of weight $1 \times W_{SFV}$. However when m = 20 in column 7, 97% of the fatigue damage occurs for vehicles of weight $6.5 \times W_{SFV}$. In other words, the weight of vehicle that causes the maximum fatigue damage depends on 'm' and this weight increases as 'm' increases. Another way of viewing this distribution of fatigue damage is to consider the effect of placing a weight restriction on the bridge that eliminates levels 1 and 2, that is the fatigue vehicles of weight $6.5W_{SFV}$ and $5W_{SFV}$ are prevented from crossing the bridge. For the welded component in which m = 3, placing the weight restriction will only reduce the fatigue damage by 2 + 5 = 7% which can be considered to be insignificant. However for stud shear connectors in which m = 5.1, the same weight restriction will reduce the fatigue damage by 25 + 32 = 57% which would significantly increase the fatigue life of the structure.

14.4 Cyclic stress resultants

14.4.1 *General*

So far in this chapter, it has been shown how the numerous fatigue vehicles that traverse the bridge can be represented by a standard fatigue vehicle and a load spectrum. The variation of the stress resultants that the standard fatigue vehicle induces on a component, as it traverses the bridge, are determined in this section using influence lines. These influence lines are then converted to equivalent ranges of cyclic forces that are used to form a *force spectrum*, that is analogous to the load spectrum already described, and which is required for the fatigue analysis procedure developed in Section 14.6

14.4.2 *Influence line diagrams*

To illustrate a simple influence line analysis, let us move the standard fatigue vehicle in Figure 14.5(a) that has an axle length of L/4 and axle weight of V across the simply supported beam in (b) of span L. We will determine how the vertical shear force varies at a design point which is located at the quarter-span at section D in (b). The variation of these vertical shear forces could be used to design the shear connection at this design point.

The procedure adopted here is to determine influence lines for each loaded axle and then to combine these influence lines to determine the influence line when the vehicle is fully loaded. The analysis when the rear axle is loaded is shown in Figures 14.5(c) to (i) where the standard fatigue vehicle is moved up to the beam in (c), in steps of L/4 along the beam in (d) to (h), and off the beam in (i). The distribution along the beam of the vertical shear force is plotted at each analysis step. From these distributions, the vertical shear force at the design point at section D can be seen. It is important that analyses are conducted when the loaded axle is just to the left of the design point and then just to the right of the design point as in (e) and (f), as this movement causes a step change in the vertical shear force.

The vertical shear force at the design point at section D of the beam in Figure 14.5(b) can be obtained from Figures 14.5(c) to (i) for different positions of the vehicle. These vertical shear forces are plotted as an influence line in Figure 14.6(b) where the front axle has been used as the reference point for the position of the vehicle. For example, when the front axle is just to the left of section E in Figure 14.5(e), then the shear force at the design point at D is 0.25V which is shown as point (1) in Figure 14.6(b). However when the front axle moves to the right of section E in Figure 14.5(f), then the vertical shear force at the design point reverses in direction to 0.75V which is shown as point (2) in Figure 14.6(b).

The same procedure has been used to determine the influence line when the front axle is loaded and the results are plotted in Figure 14.6(c). Combining (b) and (c) gives the influence line when both axles are loaded in (d), which is the variation in the vertical shear force at the quarter-span design point as the standard fatigue vehicle traverses the beam. This influence line variation will be converted in Section 14.4.3 to cyclic ranges of shear force for use in the fatigue analysis procedure that is developed in Section 14.6.

14.4.2.1 **Example 14.5** *Shear flow influence lines along length of beam*

A standard fatigue vehicle consisting of two axles is moved across a simply supported composite beam such as that shown in Figure 14.6(a). The load imposed by an axle of the standard fatigue vehicle $V = 180$ kN, the longitudinal spacing of the axles is 8 m, and the span of the beam is 32 m. The shear flow force acting at the interface between the steel and concrete components of the composite beam is given by the following well known equation which is also given in slightly different terms in Eq. 3.3.

$$q = V\left(\frac{A_c y_c}{I_{nc}}\right) = VK \qquad (14.4)$$

where V is the vertical shear force, A_c is the cross-sectional area of the concrete element, y_c is the distance between the centroid of the concrete element and the

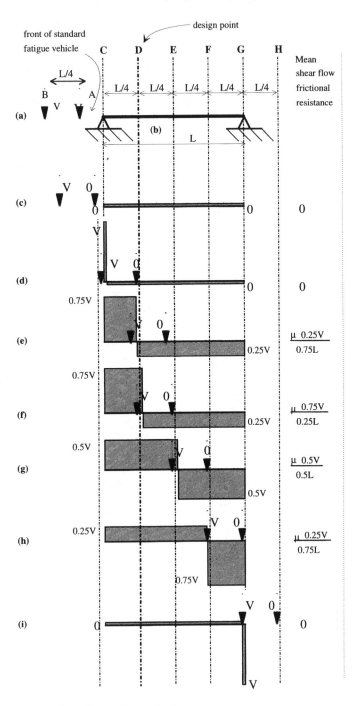

Figure 14.5 Influence line analysis at quarter span

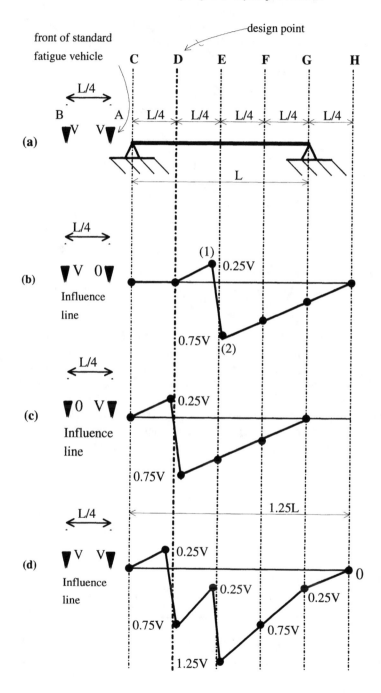

Figure 14.6 Influence line at quarter span

centroid of the transformed concrete section of the composite beam, and I_{nc} is the second moment of area of the composite section transformed to a concrete section. For the 32 m beam in this example, it will be assumed that, from the cross-section of the composite beam, the parameter in Eq. 14.4 of $A_c y_c / I_{nc} = K = 0.5 \times 10^{-3}$ mm^{-1}.

Table 14.3 Influence lines at quarter-span

Position:	C	D		E		F		G		H
(1) q (KV)	0	+0.25	-0.75	-0.25	-1.25	-0.75	-0.75	-0.25	-0.25	0
(2) q (N/mm)	0	+22.5	-67.5	-22.5	-113	-67.5	-67.5	-22.5	-22.5	0
(q_{do} without friction)										
(3) Q_{fric} (μV/L)	0	0.33	3	1.33	4	1.33	1.33	0.33	0.33	0
(4) Q_{fric} (N/mm)	0	1.3	11.8	5.3	15.8	5.3	5.3	1.3	1.3	0
(5) q_{do}(with friction)										
N/mm	0	+21.2	-55.7	-17.2	-96.7	-62.2	-62.2	-21.2	-21.2	0

As the axle spacing of the standard fatigue vehicle, in this example, is a quarter of the span of the beam, the previous analyses depicted in Figures 14.5 and 14.6 apply directly to this beam when the design point is at a quarter-span. The results of the influence line analysis in Figure 14.6(d) are listed in row 1 in Table 14.3.

Applying Eq. 14.4 with the appropriate values for K and V, gives the shear flow forces in row 2 in Table 14.3; these are the shear flow forces that the dowel action of the mechanical shear connectors have to resist q_{do} when there is no other mechanism for shear transfer such as interface friction The same analytical procedure was applied to determining the influence lines for design points at both the support at section C in Figure 14.5(b) and at mid-span at section E, and the results are given in rows 1 and 2 in Tables 14.4 and 14.5. These influence line diagrams, at all three design points, are plotted in Figure 14.7.

14.4.3   Equivalent range of cyclic forces

The influence line diagram in Figure 14.6(d) quantifies the variation of the vertical shear force at the design point but does not quantify the magnitude of the cyclic range that causes fatigue damage. Hence, the influence line diagram has to be

Table 14.4 Influence lines at support

Position:	C	D		E		F		G		H
(1) q (KV)	-1	-0.75	-1.75	-1.25	-1.25	-0.75	-0.75	-0.25	-0.25	0
(2) q (N/mm)	-90	-67.5	-158	-113	-113	-67.5	-67.5	-22.5	-22.5	0
(q_{do} without friction)										
(3) Q_{fric} (μV/L)	∞	3	∞	4	4	1.33	1.33	0.33	0.33	0
(4) Q_{fric} (N/mm)	∞	11.8	∞	15.8	15.8	5.3	5.3	1.3	1.3	0
(5) q_{do}	0	-55.7	0	-96.7	-96.7	-62.2	-62.2	-21.2	-21.2	0
N/mm (with friction)										

Table 14.5 Influence lines at mid-span

Position:	C	D		E		F		G	H
(1) q (KV)	0	+0.25	+0.25	+0.75	-0.25	+0.25	-0.75	-0.25	-0.25	0
(2) q (N/mm)	0	+22.5	+22.5	+67.5	-22.5	+22.5	-67.5	-22.5	-22.5	0
(q_{do} without friction)									
(3) Q_{fric} ($\mu V/L$)	0	0.33	0.33	1.33	1.33	1.33	1.33	0.33	0.33	0
(4) Q_{fric} (N/mm)	0	1.3	1.3	5.3	5.3	5.3	5.3	1.3	1.3	0
(5) q_{do}	0	+21.2	+21.2	+62.2	-17.2	+17.2	-62.2	-21.2	-21.2	0
N/mm (with friction)									

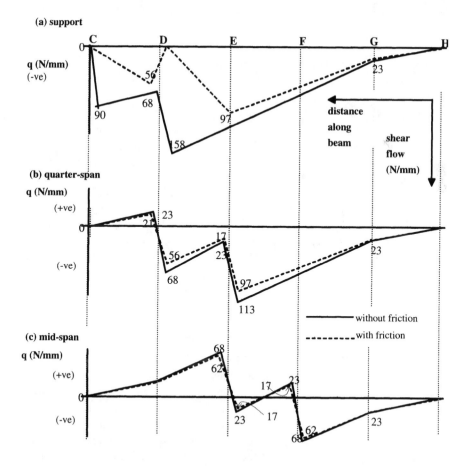

Figure 14.7 Shear flow force influence line diagrams

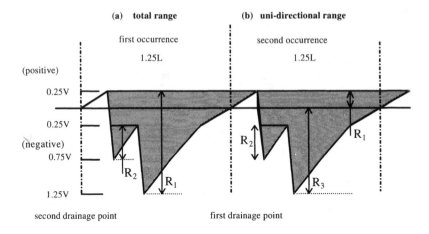

Figure 14.8 Equivalent range using reservoir method

converted to equivalent ranges of cyclic stress resultants for use in the fatigue analysis procedure developed in Section 14.6.

The reservoir approach[2] can be used to convert the influence line diagram in Figure 14.6(d) into equivalent cyclic ranges as illustrated in Figure 14.8. In this approach, two of the same influence line diagram are placed adjacent to each other as shown in (a) and (b). This diagram can now be visualized as the cross-section of a reservoir that must be emptied by successively draining from the lowest point that contains water. The procedure is repeated at the different drainage points at the parts of the reservoir that contain water until the reservoir is completely empty. The depth from the water surface at the commencement of drainage to a drainage point is an equivalent cyclic range R.

When the total range of the stress resultant is required, as in the fatigue design of stud shear connectors and other welded components, then the positive and negative shear forces have to be combined as in Figure 14.8(a). In this case, the equivalent cyclic ranges are R_1 and R_2. When the uni-directional range is required as may occur in the fatigue analysis of an unwelded steel component, then the positive and negative portions need to be considered independently as in Figure 14.8(b), where it can be seen that there are now three cyclic ranges.

14.4.3.1 **Example 14.6** *Equivalent range of cyclic forces*

Applying the reservoir approach to the influence line diagrams listed in rows 2 of Tables 14.3 to 14.5, and which are plotted in Figure 14.7, gives the cyclic shear flow ranges in rows 1 and 2 in Table 14.6 where each range occurs once, that is the frequency f = 1. It is worth noting that Figure 14.7 depicts the variation in the shear flow for unpropped construction where the steel beam resists all the dead load and the composite beam resists all the live load due to the vehicular traversals. The effect of propped construction is to maintain the same shape of the influence line diagram but to displace the origin of the ordinate. Hence propped construction will not affect the total range, as required for the analysis of the shear connection, but will affect the uni-directional range.

Also of interest is the maximum uni-directional peak shear flow that can also be derived from the influence line diagrams and which is tabulated as $(q_s)_{max}$ in row 3 in Table 14.6. This is the maximum static shear flow force imposed by the traversal of the standard fatigue vehicle and is an indication of the maximum static strength required of the shear connectors. For example, if the maximum overload W_o is a vehicle that has the same shape as the standard fatigue vehicle but is say 9 times its weight, that is $W_o = 9W_{SFV}$, then the static shear flow strength of the shear connectors q_o must be at least $9(q_s)_{max}$ as tabulated in row 4. It has been assumed in this analysis that the beam has been constructed using unpropped construction, that is the shear connectors only resist the live load due to the traversals of the vehicles. If propped construction had been used, then this maximum uni-directional shear flow would have to be superposed on that due to the dead load. A comparison of the methods of analysis for propped and unpropped construction is given in Section 3.5.

14.4.4   *Force spectrum*

From the influence line analyses, the magnitudes of the ranges of the cyclic stress resultants R and the frequencies f at which they occur can be tabulated as a force spectrum as shown in columns 2 and 3 in Table 14.7. In Section 14.2.3.1(c), it was

Table 14.6 Shear flows

(1) (N and mm)	(2) Support (N and mm)	(3) ¼ -span (N and mm)	(4) Mid-span
No Friction:			
(1) $(q_t)_1$ (f = 1)	$158 - 0 = 158$	$113 + 23 = 136$	$68 + 68 = 136$
(2) $(q_t)_2$ (f = 1)	$90 - 68 = 22$	$68 - 23 = 45$	$23 + 23 = 46$
(3) $(q_s)_{max}$	158	113	68
(4) q_o $(9(q_s)_{max})$	1422	1017	612
(5) $F_f(\times 10^9)$	163	76	76
With Friction:			
(6) $(q_t)_1$ (f = 1)	$97 + 0 = 97$	$97 + 21 = 118$	$62 + 62 = 124$
(7) $(q_t)_2$ (f = 1)	$56 + 0 = 56$	$56 - 17 = 39$	$17 + 17 = 34$
(8) $(q_s)_{max}$	97	97	62
(9) $F_f(\times 10^9)$	14	37	48

Table 14.7 Format of Force Spectrum

Level (k) (1)	Range (R) (2)	frequency (f) (3)	fR^m (4)
1	R_1	f_1	$f_1 R_1{}^m$
2	R_2	f_2	$f_2 R_2{}^m$
Z	R_Z	f_Z	$f_Z R_Z{}^m$

$$F_f = \sum fR^m$$

shown that the fatigue damage term N_k/E_k in Eqs. 14.2 and 14.3 is proportional to $N_k R^m$. As the number of cycles N of range R are also proportional to the frequency f at which they occur during the traversal of a standard fatigue vehicle, it can be seen that the parameter fR^m in column 4 in Table 14.7 is also a fatigue damage parameter. It has already been shown that the fatigue damage in a load spectrum is given by the parameter BW^m in column 4 of Table 14.1. Hence, the force spectrum in Table 14.7 is analogous to the load spectrum in Table 14.1, as both are a measure of the fatigue damage that the cyclic loads induce.

14.4.4.1 **Example 14.7** *Distribution of fatigue damage in a force spectrum*

The magnitudes of the ranges and frequencies of the cyclic loads in Figure 14.7, which were derived from the influence line analysis of the beam in Figure 14.5(b), are listed as force spectrums in Table 14.8 for design points at the support, quarter-span and mid-span. The fatigue damage term $fq_t{}^m$ has been calculated for stud shear connectors where m = 5.1. It can be seen that at each of the design points, the fatigue damage is dominated by the largest cyclic range. This is because the frequency of the smaller range is equal to that of the larger range, that is $f_1 = f_2 = 1$, so that the distribution of the fatigue damage is now directly proportional to $q_t{}^m$. This is why it is common practice to base the fatigue analysis purely on the larger cyclic range.

14.5 Frictional shear flow resistance

14.5.1 *General*

The shear flow forces on the mechanical shear connectors in a composite beam are reduced by the friction that acts at the interface between the concrete and steel components and, hence, the friction extends the fatigue life of the mechanical shear connection. A procedure is described here for quantifying the frictional shear flow resistance in a form that can be used in the fatigue analysis procedure developed in Section 14.6 and which is then used in Chapter 15 for the assessment of the strength and endurance of existing composite bridge beams.

Table 14.8 Variation in the Force Spectrum along beam

Level	Support			Quarter- span			Mid-span		
	q_t (N/mm)	f	$fq_t^{5.1}$ ($\times10^9$)	q_t (N/mm)	f	$fq_t^{5.1}$ ($\times10^9$) (N/mm)	q_t	f ($\times10^9$)	$fq_t^{5.1}$
1	158	1	163.4	136	1	76.0	136	1	76.0
2	22	1	0.007	45	1	0.27	46	1	0.30
			$\Sigma=163=F_f$			$\Sigma=76=F_f$			$\Sigma=76=F_f$

14.5.2 *Frictional resistance*

The effect of interface friction is illustrated in Figure 14.9 by considering the traversal of a point load of magnitude V across the top surface of a simply supported beam as shown in Figure 14.9(a). From the load paths shown arrowed in (a), it can be seen that normal force across the steel/concrete interface of the shear span of length L_1 is the shear force in that shear span of V_1, from which, the interface frictional resistance in this shear span of length L_1 is μV_1, where μ is the coefficient of friction between the steel and concrete components at the interface. Therefore, the mean frictional shear flow resistance in the shear span of length L_1 is $(Q_{fric})_1 = \mu V_1/L_1$. Similarly, the mean frictional shear flow resistance in the shear span of length L_2 is $(Q_{fric})_2 = \mu V_2/L_2$.

The shear flow force resisted by the dowel action of the mechanical shear connectors q_{do} is the shear flow force acting at the interface q less the mean frictional shear flow resistance Q_{fric}. The shear flow force at the steel/concrete interface q is given by Eq. 14.4, therefore, the shear flow force resisted by the dowel action of the shear connectors is given by

$$\left(q_{do}\right)_n = V_n \left\langle \frac{A_c y_c}{I_{nc}} - \frac{\mu}{L_n} \right\rangle \tag{14.5}$$

where $(q_{do})_n$ is the shear flow force on the connectors in a span designated n of length L_n, V_n is the vertical shear force in shear span n, and μ is the coefficient of friction between steel and concrete that can be taken as about 0.7. It should be noted that when the bracket in Eq. 14.5 becomes negative in theory, this means that the frictional shear flow resistance Q_{fric} is greater than the shear flow force q_{flow}. In this case, the mechanical shear connectors do not slip so that the term in the bracket in Eq. 14.5 should be equated to zero, that is $q_{do} = 0$.

Equation 14.5 can be used to derive the effect of friction on the shear flow force envelope for the point load moving across the beam in Figure 14.9(a). For example, line A_1 is the envelope for the shear flow forces q in shear span L_1 which is given by the first term in the bracket on the right hand side of Eq. 14.5 and line A_2 is the

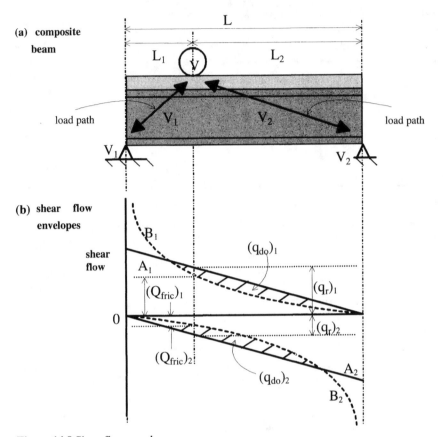

Figure 14.9 Shear flow envelopes

shear span tends to zero. To complete the picture, the shear flow frictional resistance when the design point is at section E at the mid-span in Figure 14.6(b) is given in row 3 in Table 14.5.

14.5.3.1 **Example 14.8** *Shear flow frictional resistance*
(a) Influence lines

Let us continue the analysis in Example 14.5 in Section 14.4.2.1 in which a simply supported beam of span L = 32 m was traversed by a standard fatigue vehicle that had two axles of spacing 8 m and a load per axle of 180 kN. The only additional information required, to that already given in Example 14.5, for determining the shear flow frictional resistance is the coefficient of friction which will be assumed to be $\mu = 0.7$.

For the axle spacing and beam length chosen in this problem, the shear flow frictional resistances have already been determined as a function of $\mu V/L$ in rows 3 in Tables 14.3 to 14.5. These resistances have been converted to shear flow frictional resistances in rows 4 of these tables. Rows 2 of these tables gives the shear flow force acting at the interface.

envelope in shear span L_2. Line B_1 is the shear flow frictional resistance Q_{fric} in shear span L_1 which is given by the second term in Eq. 14.5. Therefore, the hatched region is the shear flow force q_{do} acting on the connectors. The sum of both components of the hatched regions, that is $(q_{do})_1 + (q_{do})_2$, is the total range of the cyclic shear flow force acting on the mechanical shear connectors. It can be seen that the frictional resistance has the greatest benefit adjacent to the supports where in this case $q_{do} = 0$ and has the least effect at mid-span where q_{do} is at its maximum.

14.5.3 *Frictional resistance influence line diagrams*

The shear flow frictional resistance can be derived from the influence line analyses that are used to determine the ranges of shear flows at a design point, such as the analysis illustrated in Figure 14.5 that has already been described in detail in Section 14.4.2. Take for example the distribution of vertical shear in Figure 14.5(e): the design point occurs in the right hand side shear span D-G of length 0.75L; the vertical shear force at the design point is 0.25V which is the normal force across the interface along the right hand shear span; hence, the interface frictional force is $\mu 0.25V$; and as the interface frictional force acts over a shear span of length 0.75L, the mean shear flow frictional resistance is $\mu 0.25V/0.75L$ as listed on the right hand side of the figure. When the rear axle is moved to the right of the design point as in (f): the design point now lies along the left hand shear span of length 0.25L; the vertical shear force at the design point is now 0.75V; so that the mean shear flow frictional resistance at the design point is $\mu 0.75V/0.25L$.

It is worth emphasizing the point that the shear flow frictional resistance is not dependent on the sign of the vertical shear force, as the shear flow frictional resistance simply resists movement. A convenient approach in the analysis is to attach the appropriate sign of the vertical shear force to V_n in Eq. 14.5. The terms within the bracket now depend only on the geometric and material properties of the beam and, as a further reminder, when this bracket is negative it should be assumed to be zero. The shear flow frictional resistances listed on the right hand side of Figure 14.5 are plotted as an influence line diagram in Figure 14.10(a) using the front axle as the reference point for the position of the vehicle on the beam. The results marked (1) and (2) in Figure 14.10(a) correspond to the analyses in Figure 14.5(e) and (f) where $\mu \times 0.25V/0.75L = 0.33\mu V/L$, and $\mu \times 0.75V/0.25L = 3\mu V/L$. It can be seen that even though the sign of the vertical shear force at the design point changed in the analyses in Figures 14.5(e) and (f), both frictional shear flow resistances are shown as positive in the influence line diagram in Figure 14.10(a).

The shear flow frictional resistance for when the rear axle is loaded has been derived using the same procedure and is shown in Figure 14.10(b). Adding (a) to (b) gives the shear flow frictional resistance in (c) when both axles are loaded and which applies to the design point at a quarter-span. The results are tabulated as Q_{fric} in row 3 in Table 14.3. The shear flow frictional resistance when the design point is at the support at section C in Figure 14.6(b) is listed in row 3 in Table 14.4 and plotted in Figure 14.11. It can be seen that the frictional resistance tends to infinity when the axle loads are adjacent to the design point at the supports because the length of the

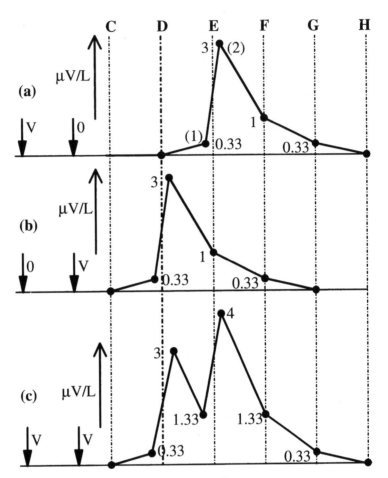

Figure 14.10 Frictional resistance influence line at quarter-span

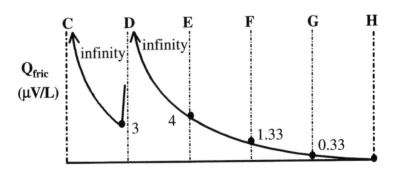

Figure 14.11 Frictional resistance influence line at support

Reducing the magnitude of rows 2 in these tables by the magnitude of rows 4 of these tables gives row 5, which is the shear flow force acting on the mechanical shear connectors. The results are plotted in Figure 14.7 as the broken line labelled 'with friction'.

The differences in Figure 14.7 between the shear flows 'without friction' and the shear flows 'with friction' are the reductions due to friction of the shear flow forces resisted by the mechanical shear connectors. It can be seen that the effect of friction is large at the supports but relatively small at mid-span.

(b) Equivalent range of cyclic shear flow forces

Applying the reservoir technique described in Section 14.4.3, to the influence line diagrams 'with friction' in Figure 14.7, gives the results in rows 6 to 8 in Table 14.6. Let us compare these results 'with friction' to the results in rows 1 to 3 in which there was no friction. At the supports, the beneficial effect of friction has reduced the maximum cyclic range from 158 to 97 N/mm that is by 39% and the peak uni-directional load by the same amount, which means that the shear connectors at the supports can resist a higher static load and will last much longer than originally anticipated. At mid-span, the maximum range and peak uni-directional load has only reduced from 136 to 124 N/mm that is by 9% and, hence, at mid-span the beneficial effect of friction is fairly minor.

14.6 Generic fatigue equation

14.6.1 *General*

So far in this chapter, we have discussed general methods for representing: the fatigue material properties; the fatigue vehicular loads; and the cyclic stress resultants that induce fatigue failure. A generic fatigue equation will now be developed that incorporates all these general representations and which will be used in Chapter 15 for the design and assessment of the shear connection composite bridge beams.

14.6.2 *Generic fatigue material properties*

Both the crack propagation accumulated damage law of Eq. 14.2 and the crack initiation accumulated damage law of Eq. 14.3 can be written in the following generic form.

$$\sum \frac{N}{E} = A \tag{14.6}$$

The general form of the endurance given by Eq. 14.1 can be represented by the following generic form.

$$E = C \left(\frac{R}{X} \right)^{-m} \tag{14.7}$$

It is often convenient in fatigue analyses to represent the cyclic stress resultant as the ratio R/X in Eq. 14.7. For example in the design of a fillet weld, R could be the shear flow force derived from the analysis of the beam and X the unknown transverse width of the weld to be determined in the design, so that R/X is the shear stress τ in a design procedure where the endurance of the fillet weld is given in terms of the shear stress τ. However in the assessment of an existing weld of known dimensions, R could represent the shear stress τ so that $X = 1$.

14.6.3 *Fatigue damage analysis*
14.6.3.1 Cyclic stress resultants

The generic fatigue equation will be derived by considering the following very simple fatigue problem: a composite beam is subjected to a total of T traversals of fatigue vehicles; the fatigue vehicles can be represented by one standard fatigue vehicle; there are only two weights of fatigue vehicles as listed in Table 14.9; the traversal of the standard fatigue vehicle produces two ranges of the stress resultant as listed in Table 14.10.

From Tables 14.9 and 14.10, it can be deduced that there are only four ranges of the stress resultants which are: W_1R_1 that occurs B_1Tf_1 times; W_1R_2 that occurs B_1Tf_2 times; W_2R_1 that occurs B_2Tf_1 times; and W_2R_2 that occurs B_2Tf_2 times.

14.6.3.2 *Fatigue damage*

Let us consider the fatigue damage due to the first cyclic range W_1R_1 from Section 14.6.3.1 which in terms of the parameter X can be written as W_1R_1/X. Substituting this range into the generic endurance of Eq. 14.7 gives the endurance for the first of the cyclic range W_1R_1 as

$$E_{first} = CW_1^{-m} R_1^{-m} X^{m} \qquad (14.8)$$

Table 14.9 Load spectrum with two weights of vehicles

(1) Level (x)	(2) Weight (W)	(3) Probability (B)	(4) B W^m
1	W_1	B_1	$B_1W_1{}^m$
2	W_2	B_2	$B_2W_2{}^m$
		$\Sigma = 1$	$L_f = \Sigma BW^m$

Table 14.10 Force spectrum with two ranges

(1) Level (k)	(2) Range (R)	(3) frequency (f)	(4) f R^m
1	R_1	f_1	$f_1R_1{}^m$
2	R_2	f_2	$f_2R_2{}^m$
			$F_f = \Sigma fR^m$

The number of cycles of the first range W_1R_1 from Section 14.6.3.1 has been shown to be $N_{first} = B_1Tf_1$. Substituting the values for N_{first} and E_{first} into the generic accumulated damage law of Eq. 14.6 gives the fatigue damage A_{first} due do the first of the cyclic ranges W_1R_1 as

$$A_{first} = C^{-1}(B_1W_1{}^m)(f_1R_1{}^m)TX^{-m} \qquad (14.9)$$

The same procedure can be applied to the remaining three cyclic ranges of W_1R_2, W_2R_1 and W_2R_2 to give the following fatigue damage terms.

$$A_{second} = C^{-1}(B_1W_1{}^m)(f_2R_2{}^m)TX^{-m} \qquad (14.10)$$

$$A_{third} = C^{-1}(B_2W_2{}^m)(f_1R_1{}^m)TX^{-m} \qquad (14.11)$$

$$A_{third} = C^{-1}(B_2W_2{}^m)(f_2R_2{}^m)TX^{-m} \qquad (14.12)$$

Summing the fatigue damage in Eqs. 14.9 to 14.12, as required in Eq. 14.6, gives the total fatigue damage 'A' as

$$A = T(B_1W_1{}^m + B_2W_2{}^m)(f_1R_1{}^m + f_2R_2{}^m)X^{-m}C^{-1}$$

$$\qquad (14.13)$$

14.6.4 *Generic fatigue equation*
14.6.4.1 Load constant
The parameter $(B_1W_1{}^m + B_2W_2{}^m)$ in Eq. 14.13 is the sum of the fatigue damage terms in column 4 of the load spectrum in Table 14.9. This fatigue parameter will be referred to as the *load constant* and will be denoted by the symbol L_f. Hence, the load constant L_f is given by the following equation

$$L_f = \sum_{x=1}^{x=i}(B_xW_x{}^m) \qquad (14.14)$$

where there are 'i' levels in the load spectrum as in Table 14.1. The derivation of L_f is also shown in column 4 of Table 14.1.

14.6.4.2 *Force constant*
The fatigue parameter $(f_1R_1{}^m + f_2R_2{}^m)$ in Eq. 14.13 will be referred to as the *force constant* F_f and can be obtained from the force spectrum in Table 14.10. Hence, this fatigue parameter can be derived from the following equation

$$F_f = \sum_{}^{k=z}(f_kR_k{}^m) \qquad (14.15)$$

$k = 1$

where there are z levels in the force spectrum as shown in Table 14.7 and where the force constant is also derived in column 4 of this table.

14.6.4.3 *Generic fatigue equation*

Substituting the fatigue damage terms L_f and F_f in Eqs. 14.14 and 14.15 into Eq. 14.13 gives the following generic fatigue equation.

$$X^{-m} = \frac{AC}{\sum\limits_{y=1}^{y=j} \left(TF_f L_f\right)_y} = \frac{AC}{T_1\left(F_f\right)_1\left(L_f\right)_1 + T_2\left(F_f\right)_2\left(L_f\right)_2 \cdots + T_j\left(F_f\right)_j\left(L_f\right)_j} \quad (14.16)$$

The parameters TFL in the denominator of Eq. 14.16 quantifies the fatigue damage that the component is subjected to. For convenience in the fatigue analysis, we will define a *fatigue zone* as a period of T traversals of fatigue vehicles during which both F_f and L_f are constant. As it has been shown in Section 14.2.3.1 that the accumulated damage laws are based on a linear variation in the residual strength, the sequence at which the fatigue zones occur does not affect the overall damage[1], so that the fatigue damage due to each fatigue zone can be summed as shown in Eq. 14.16 where there are j fatigue zones.

14.6.4.1 **Example 14.9** *Load constant and force constant*

Examples of the derivation of the load constant L_f are given in Table 14.2 for different values of the fatigue exponent m. It is fairly obvious that L_f increases as m increases. The variation in the force constant fatigue damage term F_f is shown in Table 14.6 for the beam analysed in Example 14.5 in Section 14.4.2.1. The beneficial effect of friction is ignored in the fatigue design of new composite beams and, hence, the results in row 5 of Table 14.6 would be used in the design of new bridges. It can be seen that the greatest fatigue damage along the length of the beam occurs to the shear connectors adjacent to the supports where $F_f \propto 163$.

In contrast to the design of new structures in the previous paragraph, in the fatigue assessment of an existing composite beam it may be a requirement to determine a realistic residual strength and residual endurance. In this case the effects of friction could be included as in the analyses in row 9 of Table 14.6. It can be seen in row 9 that friction at the support has reduced the fatigue term by an order of magnitude from $F_f \propto 163$ to $F_f \propto 14$. Furthermore, the greatest fatigue damage now occurs adjacent to the mid-span where $F_f \propto 48$.

14.7 References

1. Oehlers, D.J. and Bradford, M.A. (1995). Composite Steel and Concrete Structural Members: Fundamental Behaviour. Pergamon Press, Oxford.
2. B.S.I. (1980). BS5400 : Part 10 : 1980. Steel, concrete and composite bridges, Part 10. Code of practice for fatigue. British Standards Institution, London.

15 Fatigue analysis of stud shear connections

15.1 Introduction

The fatigue behaviour of composite steel and concrete beams with mechanical stud shear connectors is unique. The concrete slab of the composite beam exhibits the usual well known time dependent characteristics of creep and shrinkage. However, tests have shown conclusively that the shear connection reduces in strength and stiffness immediately cyclic loads are applied, which means that both the degree of interaction and the degree of shear connection of the composite bridge beam changes each time a fatigue vehicle traverses the bridge. Hence, the strength and stiffness of an existing composite beam that is subjected to fatigue loads is continually changing with time, which makes the fatigue design and assessment a very interesting problem.

In Chapter 14, standard procedures were described for determining the cyclic stress resultants that cause fatigue damage, and for using these stress resultants to determine the residual strength and endurance of a component of a structure. These standard procedures were described in a form that can be applied in theory to any type of stress resultant and any type of structural component. This chapter will concentrate on the fatigue design and assessment of the stud shear connectors in composite bridge beams, and will deal with the shear flow forces that act on stud shear connectors q and the shear flow strengths Q that they require for strength and endurance. It is worth reiterating that even though the fatigue limit state is an ultimate limit state, the analysis is elastic, in contrast to the ultimate limit state of strength that is often governed by the upper bound rigid plastic analysis.

The crack initiation and crack propagation fatigue material properties of stud shear connectors are first described in Section 15.2. This is then followed in Section 15.3 by the details of a composite beam that are used in Sections 15.4 and 15.5 to illustrate the crack initiation approach and crack propagation approach of the fatigue design of new bridge beams, and the fatigue assessment of existing bridge beams. Finally in Section 15.6, the crack propagation fatigue approach is applied to assessing the strength and endurance of composite beams in buildings that are subjected to cyclic loads.

15.2 Stud shear connector fatigue material properties

Endurance equations for stud shear connectors have the general form shown in Eq. 14.1 and can be categorized in terms of whether they follow the crack initiation approach described in Section 14.2.3.1(b) or the crack propagation approach described in Section 14.2.3.1(a).

15.2.1 *Crack initiation properties*

The following two equations are examples[1] of crack initiation fatigue endurance equations for stud shear connectors.

$$E_{ch} = 10^{\left(15.9-\frac{0.7}{\sqrt{n}}\right)}\left(\frac{R}{A_{sh}}\right)^{-5.1} \tag{15.1}$$

where the units and in N and mm, and

$$E_{ch} = 10^{\left(2.27-\frac{0.70}{\sqrt{n}}\right)}\left(\frac{R}{D_{max}}\right)^{-5.4} \tag{15.2}$$

where E_{ch} refers to the characteristic endurance at two standard deviations, n is the number of connectors in a group within the composite beam that can be assumed to fail together, R is the magnitude of the range of a cyclic shear load acting on a stud shear connector, A_{sh} is the cross-sectional area of the shank of the stud shear connector, and D_{max} is the static strength of the stud shear connector, that is its strength prior to cyclic loading.

The fatigue parameter that controls the endurance is often assumed to be R/A_{sh} as shown in Eq. 15.1 The fatigue term R/A_{sh} is a pseudo shear stress that is supposed to be acting on the shank of the stud shear connector; it is a pseudo shear stress as it ignores the shear forces resisted by the weld-collar of the stud shear connector as shown in Figure 5.1. However, this approach allows stud shear connectors to be designed using the same standard procedures as are used for other welded metal components. It should be remembered that endurance equations of this form are dimensionally incorrect, that is the constant of the equation depends on the units being used and, hence, the requirement of N and mm for Eq. 15.1. Another common form of the endurance equation is shown in Eq. 15.2 where the fatigue parameter now depends on R/D_{max}. This form of endurance equation is dimensionally correct and published statistical analyses[1] of fatigue endurances have shown that the fatigue parameter R/D_{max} gives the least scatter of results.

In the analysis of composite beams it is often more convenient to deal with shear flow forces and shear flow strengths as these are derived directly from the analysis of the composite beam, and are not dependent on the choice of the mechanical shear connector. In which case, Eq. 15.2 can be written in the following form

$$E_{ch} = 10^{\left(2.27-\frac{0.70}{\sqrt{n}}\right)}\left(\frac{q_t}{Q_D}\right)^{-5.4} \tag{15.3}$$

where q_t is the range of the cyclic shear flow force, and Q_D is the shear flow strength of the stud shear connectors prior to cyclic loads, that is the static strength D_{max} of the stud shear connectors per unit length.

The accumulated damage law for the crack initiation procedure is given by the following equations

$$\sum_{k=1}^{k=z} \frac{N_k}{E_k} \leq 1 \tag{15.4a}$$

$$E_k = f(q_t) \tag{15.4b}$$

where there are z magnitudes of the cyclic shear flow ranges q_t, N_k applications of range $(q_t)_k$, and in which the shear connection when subjected to a range of magnitude $(q_t)_k$ has an endurance E_k which is a function of q_t such as that given by Eq. 15.3.

15.2.2 Crack propagation properties

The crack propagation fatigue endurance[1] for stud shear connectors is given by

$$E_{ch} = 10^{\left(3.12 - \frac{0.70}{\sqrt{n}}\right)} \left(\frac{R}{D_{max}}\right)^{-5.1} \tag{15.5}$$

that can be written in the following form in terms of the shear flows

$$E_{ch} = 10^{\left(3.12 - \frac{0.70}{\sqrt{n}}\right)} \left(\frac{q_t}{Q_D}\right)^{-5.1} \tag{15.6}$$

for which the following accumulated damage law is applicable

$$\sum_{k=1}^{k=z} \frac{N_k}{E_k} \leq 1 - \frac{q_o}{Q_D} \tag{15.7a}$$

$$E_k = f(q_t) \tag{15.7b}$$

where q_o is the maximum uni-directional shear flow force that the stud shear connection has to resist, which could be the maximum uni-directional shear flow force induced by the maximum overload that the composite beam has to resist, and Eq. 15.7b has been included to remind the reader that the endurance E_k is a function of the cyclic shear flow range q_t.

15.3 Details of composite beam

The composite bridge beam that was analysed in detail in Chapter 14 will also be used in this chapter to illustrate the design and assessment procedures in Sections 15.4 and 15.5. The details of the beam and the results of the analyses in Chapter 14 that are required in this chapter are summarized in this section.

The simply supported composite beam in Figure 15.1 has a span of L = 32 m, the cross-section of the composite beam when transformed to concrete has a value of K

$= A_c y_c / I_{nc} = 0.5 \times 10^{-3}$ mm^{-1}, and the coefficient of friction between the concrete slab and steel beam has a value of $\mu = 0.7$. The beam has been analysed at the three design points shown in Figure 15.1, that is at the support, at the quarter-span and at the mid-span. It will be assumed that the beam has been constructed using unpropped construction, so that the stud shear connectors only resist the live load imposed by the fatigue vehicles.

The standard fatigue vehicle that is used to represent the fatigue vehicles that cross the beam has a weight $W_{SFV} = 320$ kN, and consists of two axles that have a spacing of $L/4 = 8$ m and which are equally loaded at $V = 160$ kN. The load spectrum associated with this standard fatigue vehicle is shown in columns 2 and 3 in Table 15.1 and the load spectrum has a value of the *load constant* of $L_f = 1.14$ as shown in column 4.

The force spectrums at each design point that was derived from the longitudinal traversal of the standard fatigue vehicle across the beam is shown in Table 15.2. The results in the 'No friction' part are the shear flow forces at the interface, all of which has to be resisted by the mechanical shear connectors, as the beneficial effect of friction is ignored in this analysis. In contrast, the results in the 'With friction' part are the shear flow forces acting on the stud shear connectors, that allows for the beneficial effect of friction. The force constants F_f for each force spectrum are also listed in Table 15.2. The values in Table 15.2 are slightly different from the results of the original analyses in Table 14.8 due to rounding off errors that occurred in demonstrating the technique in Chapter 14.

The maximum uni-directional shear flow forces $(q_s)_{max}$ acting on the stud shear connectors when the standard fatigue vehicle traverses the beam are given in rows 1 and 3 in Table 15.3. It will be assumed that the maximum overload is a vehicle with the same shape of axle configuration as the standard fatigue vehicle, and that the weight of this overload vehicle $W_o = 9W_{SFV} = 2880$ kN. Hence, the maximum uni-directional shear flow force that has to be resisted by the stud shear connectors q_o is 9 times that induced by the standard fatigue vehicle and which is given in rows 2 and 4. It needs to be emphasized that this is the maximum uni-directional shear flow force when unpropped construction was used, otherwise the connectors would be subjected to an additional shear flow force due to the dead weight of the structure acting on the composite beam. Hence if propped construction had been used, then the uni-directional shear flow due to the dead load would have to be superposed on that due to the live load to obtain q_o as explained in Section 3.5.

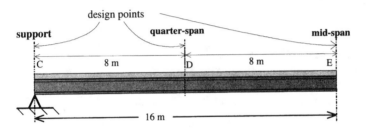

Figure 15.1 Composite beam used in the analyses

Table 15.1 Load spectrum

(1) Level	(2) W	(3) B	(4) BW^5.1	(5) W	(6) B	(7) BW^5.1	(8) W	(9) B	(10) BW^5.1
1	6.5	0.00002	0.280	6.5	0.00002	0.280	-	-	-
2	5.0	0.00010	0.367	5.0	0.00010	0.367	-	-	-
3	2.0	0.01000	0.343	2.0	0.01000	0.343	2.0	0.01000	0.343
4	1.0	0.13988	0.140	1.5	0.13988	1.106	1.5	0.13990	1.106
5	0.5	0.25000	0.007	0.5	0.25000	0.007	0.5	0.25003	0.007
6	0.2	0.60000	0.000	0.2	0.60000	0.000	0.2	0.60007	0.000
		$\Sigma B = 1$	$L_f = 1.14$		$\Sigma B = 1$	$L_f = 2.10$		$\Sigma B = 1$	$L_f = 1.46$

Table 15.2 Force Spectrum for m = 5.1

Level	Support			Quarter- span			Mid-span		
	(1) q_t (N/mm)	(2) f	(3) $fq_t^{5.1}$ ($\times 10^9$)	(4) q_t (N/mm)	(5) f	(6) $fq_t^{5.1}$ ($\times 10^9$)	(7) q_t (N/mm)	(8) f	(9) $fq_t^{5.1}$ ($\times 10^9$)
Part 1: No friction ($\mu = 0$):									
1	158	1	163.4	135	1	73.2	135	1	73.2
2	23	1	0.009	45	1	0.27	45	1	0.27
			$F_f = 163$			$F_f = 73$			$F_f = 73$
Part 2: With friction ($\mu = 0.7$):									
1	97	1	13.6	118	1	36.9	124	1	47.5
2	56	1	0.82	39	1	0.13	34	1	0.65
			$F_f = 14$			$F_f = 37$			$F_f = 48$

15.4 Crack initiation approach

In this section, the generic fatigue equation of Eq. 14.16 is developed specifically for the crack initiation approach for stud shear connectors. It is written in both a design mode and an assessment mode.

15.4.1   *Design mode*
15.4.1.1 Crack initiation fatigue design equations

Equation 14.16 was derived from the generic form of the accumulated damage law in Eq. 14.6 and the generic form of the endurance Equation 14.7. Comparing Eqs.

Table 15.3 Uni-directional shear flow forces

(1)	(2) Support (N and mm)	(3) ¼ -span (N and mm)	(4) Mid-span (N and mm)
Part 1: No Friction			
(1) $(q_s)_{max}$	158	113	68
(2) $q_o = 9(q_s)_{max} = Q_o$	1418	1013	608
Part 2: With Friction:			
(3) $(q_s)_{max}$	97	97	62
(4) $q_o = 9(q_s)_{max}$	870	870	560

14.6 and 14.7 with their crack initiation counterparts for stud shear connectors in Eqs. 15.4 and 15.3, it can be seen that generic fatigue equation of Eq. 14.16 can be written in the following form.

$$Q_f = \left[\frac{AC}{\sum\limits_{y=1}^{y=j} \left(TF_f L_f\right)_y} \right]^{-\frac{1}{5.4}} = \left[\frac{10^{\left(2.27 - \frac{0.70}{\sqrt{n}}\right)}}{\sum\limits_{y=1}^{y=j} \left(T F_f L_f\right)_y} \right]^{-\frac{1}{5.4}} \quad (15.8)$$

where Q_f is the shear flow strength of the stud shear connectors that depends purely on the fatigue endurances, and $F_f = f(q_t)$ that is the force constant is to be derived from the cyclic shear flow forces q_t.

15.4.1.2 **Example 15.1** *Fatigue design based on mean properties*

Even though the crack initiation approach does not allow directly for the gradual reduction in the strength of stud shear connectors during cyclic loading, this approach is commonly used in practice as factors of safety and the use of characteristic properties can cater indirectly for reductions in strength.

The composite beam in Figure 15.1 will be designed for T = 300 million traversals of fatigue vehicles that occurs over a period of 100 years. In Eq. 15.8, the exponent m = 5.4 and this has been used to derive both the force constants in Table 15.4, at the various design points, and the load constant L_f = 1.654 in column 4 in Table 15.5. Furthermore, it will be assumed in Eq. 15.8 that n → ∞, that is we will designing using the mean material properties. Again units of N and mm are used throughout this analysis and also in the ensuing analyses unless stated otherwise.

(a) Shear flow strengths

At the support design point in Figure 15.1, the variables in Eq. 15.8 are: $T = 300 \times 10^6$; $L_f = 1.654$ from column 4 in Table 15.5; $F_f = 746 \times 10^9$ from column 3 in the no friction part in Table 15.4; and $n \to \infty$. From which $(Q_f)_{support} = 2447$ N/mm which is shown as point A in Figure 15.2. Applying the same analysis at the quarter-span and mid-span design points in Figure 15.1, where the only change in the variables is $F_f = 320 \times 10^9$ from columns 6 and 9 in Table 15.4, gives $(Q_f)_{1/4\text{-span}} = (Q_f)_{mid\text{-span}} = 2092$ N/mm that are shown as points B and C in Figure 15.2. Also plotted as A", B" and C" in Figure 15.2 are the shear flow strengths Q_o that are required to resist the maximum overload that are listed in row 2 in Table 15.3.

Table 15.4 Force Spectrum for m = 5.4

Level	Support			Quarter- span			Mid-span		
	(1) q_t (N/mm)	(2) f	(3) $fq_t^{5.4}$ (×10⁹)	(4) q_t (N/mm)	(5) f	(6) $fq_t^{5.4}$ (×10⁹)	(7) q_t (N/mm)	(8) f	(9) $fq_t^{5.4}$ (×10⁹)
Part 1: No friction ($\mu = 0$)									
1	158	1	746.0	135	1	319.0	135	1	319.0
2	23	1	0.002	45	1	0.85	45	1	0.85
			$F_f = 746$			$F_f = 320$			$F_f = 320$
Part 2: With friction ($\mu = 0.7$)									
1	97	1	53.5	118	1	154.2	124	1	201.6
2	56	1	2.8	39	1	0.39	34	1	0.19
			$F_f = 56$			$F_f = 155$			$F_f = 202$

Table 15.5 Load Spectrum with m = 5.4

(1) Level	(2) W	(3) B	(4) BW⁵·⁴	(5) W	(6) B	(7) BW⁵·⁴
1	6.5	0.00002	0.491	6.5	0.00002	0.491
2	5.0	0.00010	0.595	5.0	0.00010	0.595
3	2.0	0.01000	0.422	2.0	0.01000	0.422
4	1.0	0.13988	0.140	1.5	0.13988	1.249
5	0.5	0.25000	0.006	0.5	0.25000	0.006
6	0.2	0.60000	0.000	0.2	0.60000	0.000
		$\Sigma B = 1$	$L_f = 1.654$		$\Sigma B = 1$	$L_f = 2.763$

It can be seen in Figure 15.2 that, in this example, Q_f is greater than Q_o throughout the length of the beam and, hence, the endurance controls the design of this beam throughout its length. The relative positions of the lines Q_f and Q_o depend on T and W_o. For example, doubling Q_o as shown by the line marked $2Q_o$ would make the static strength control the design over about half of the beam that is adjacent to the supports. Conversely reducing T would lower Q_f.

(b) Distribution of stud shear connectors

Let us assume that 19×100 mm studs are being used in this composite beam, that these shear connectors have the strengths given by Eq. 5.4, and that inserting the appropriate material properties of the stud shear connections into Eq. 5.4 gives the static or dowel strength as

$$D_{max} = 25000\left(4.7 - \frac{1.2}{\sqrt{n}}\right) \qquad (15.9)$$

where the strength is in N.

When $n \to \infty$ in Eq. 15.9, the mean dowel strength of the stud shear connector is $(D_{max})_{mean} = 117.5$ kN. The longitudinal spacing of the stud shear connectors, if they were placed in one single line, is $L_{si} = D_{max}/Q_f$, that at the supports comes to $(L_{si})_{support} = 117500/2447 = 48$ mm and at the other design points has the value $(L_{si})_{1/4-span} = (L_{si})_{mid-span} = 117500/2092 = 56$ mm.

15.4.1.3 **Example 15.2** *Characteristic material properties*

(a) Number of connectors n that fail as a group

Stud shear connectors fail as a group both because they have a ductile plateau as described in Chapter 5 and shown in Figure 5.2 and because of the incremental set[2],

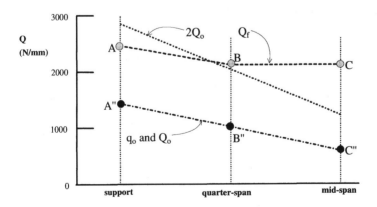

Figure 15.2 Shear flow strengths from crack initiation approach

that is the permanent set that occurs during cyclic loading. Hence, their characteristic strength and endurance is a function of the number of connectors n that can be assumed to fail as a group, as given by Eqs. 15.8 and 15.9.

In the ultimate strength design of composite beams in Chapter 4 it was a simple matter to define n as the number of connectors in a shear span, that is between the position of maximum moment and the support, as all of these connectors are required to resist the thrust at the ultimate load. Defining n in the fatigue design of an 'elastic' composite bridge beam is much more difficult as there are generally more connectors than required for equilibrium and the extent of the redistribution of shear due to the incremental set has not been quantified. It will be assumed in this analysis that only the connectors that are spread over a length equal to twice the depth of the beam, $L_{sp} = 2D$, can be assumed to fail as a group. It will be assumed that $L_{sp} = 2D = 3$ m for the beam being analysed.

(b) Spacing of connectors as a function of their characteristic strengths and endurances

As both the characteristic endurance and characteristic static strength depend on n in Eqs 15.8 and 15.9, an iterative procedure will have to be used to quantify n. Let us start by determining the spacing of the connectors at the quarter-span and mid-span regions where F_f is the same. This analysis uses the same variables as in Example 15.1 except that n now varies. From Eqs. 15.8 and 15.9 with $n \rightarrow \infty$, $Q_f = 2092$ N/mm and $D_{max} = 117{,}500$ N, from which $n = Q_f L_{sp}/D_{max} = 53.4$ mm for $L_{sp} = 3000$ mm. Now repeating the analyses with $n = 53.4$ in Eqs. 15.8 and 15.9 gives $Q_f = 2179$ N/mm and $D_{max} = 13{,}379$ N from which $n = Q_f L_{sp}/D_{max} = 57.7$. Continuing with $n = 57.7$ gives $Q_f = 2176$ N/mm, $D_{max} = 113{,}550$ N from which $n = Q_f L_{sp}/D_{max} = 57.5$. It can be seen that the number of connectors n converges rapidly. The spacing required is, therefore, $L_{si} = 3000/57.5 = 52.2$ mm. A similar iterative analysis at the supports gives $L_{si} = 44.9$ mm.

(c) Sensitivity of design to n

The variation of the longitudinal spacing L_{si} with the number of connectors that can be assumed to fail as a group n is shown in Figure 15.3. When $n = 1$ in Eqs. 15.8 and 15.9, we are dealing with the characteristic endurance and strength of an individual connector, this gives a longitudinal spacing of $L_{si} = 32$ mm at $n = 1$ as shown. When $n \rightarrow \infty$, $L_{si} \rightarrow 56$ mm which is shown as the asymptote in Figure 15.3. It can be seen in Figure 15.3 that as n increases L_{si} rapidly converges to the asymptote. It can also be seen that when n exceeds about 40 then there is very little change in L_{si}. Therefore, it is suggested that, as in most composite bridge beams where n is fairly large, the design is based on the mean endurance and strength as given in Example 15.1.

15.4.2   *Assessment mode*
15.4.2.1 Fatigue assessment equation
In the assessment of an existing composite bridge beam, the shear flow strength of the shear connection Q_f is already known, so that the variable to be determined is the residual endurance

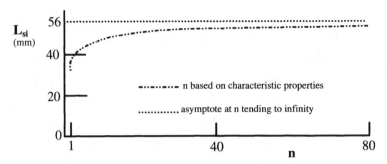

Figure 15.3 Number of connectors that fail as a group

that can be measured in terms of the remaining fatigue vehicle traversals T. Therefore, Eq. 15.8 can be rearranged into the following form that is convenient for assessment.

$$\left(TF_f\,L_f\right)_1 + \left(TF_f\,L_f\right)_2 + \ldots + \left(TF_f\,L_f\right)_j = 10^{\left(2.27 - \frac{0.70}{\sqrt{n}}\right)} Q_f^{5.4} \qquad (15.10)$$

where there are j fatigue zones (as defined in Section 14.6.4.3 as when F_f and L_f are constant), and only one unknown value of T in these j fatigue zones.

15.4.2.2 **Example 15.3** *Fatigue assessment of an existing composite bridge beam*

Suppose that the composite beam that was designed in Example 15.1 for a total of 300 million fatigue vehicle traversals has now been subjected to $T_1 = 150 \times 10^6$ fatigue vehicle traversals. Furthermore that during this fatigue zone of $T_1 = 150 \times 10^6$, the load spectrum has remained constant so that the load constant has the original design value of $(L_f)_1 = 1.654$, and the force spectrums has also remained constant so that the force constant remains at $(F_f)_1 = 320 \times 10^9$ at both mid and quarter spans and at $(F_f)_1 = 746 \times 10^9$ at the supports. Let us now assume that the weights of the commercial vehicles have been allowed to increase which has caused level 4 in column 2 in Table 15.5 to increase from $W_{FV} = 1 \times W_{SFV}$ to $W_{FV} = 1.5 \times W_{SFV}$ as shown in column 5. This has increased the load constant to $(L_f)_2 = 2.736$ as shown in column 7. It is required to determine the fatigue effect of increasing the commercial vehicle weights.

Let us at first apply Eq. 15.10 at the quarter-span and mid-span design points where the force constant F_f is the same. From Example 15.1, $Q_f = 2092$ N/mm. In the first fatigue zone, $T_1 = 150 \times 10^6$, $(F_f)_1 = 320 \times 10^9$ and $(L_f)_1 = 1.654$. In the remaining fatigue zone, the force constant remains unchanged at $(F_f)_2 = 320 \times 10^9$, as the standard fatigue vehicle and the cross-sectional properties of the bridge are unchanged, however, the load constant has increased to $(L_f)_2 = 2.736$. Hence, the only unknown in Eq. 15.10 is the number of fatigue vehicle traversals T_2 in the second fatigue zone which comes to 91 million traversals. As the bridge was originally designed for 300 million traversals over 100 years and as the number of traversals to cause failure has reduced to 150 + 91 = 241 million. The bridge will last $(241/300) \times 100 = 80$ years in total or a further 30 years, from the increase in the commercial vehicle weights, assuming the rate of loading remains unchanged. A similar analysis applied to the supports will give exactly the same result because the same change in L_f has occurred at all the design points.

15.4.2.3 **Example 15.4** *Fatigue assessment with friction*

If the endurances in Example 15.3 are not considered to be satisfactory, then a more accurate analysis would be to include the effect of friction in applying Eq. 15.10. The force constants with friction are given in Table 15.4. At the mid-span: $Q_f = 2092$ N/mm from Example 15.1; the first fatigue zone is given by $T_1 = 150 \times 10^6$, $(F_f)_1 = 202 \times 10^9$ from column 8 in the friction part of Table 15.4; and $(L_f)_1 = 1.654$ from column 4 of Table 15.5. In the second fatigue zone the force, $(F_f)_2 = 202 \times 10^9$ and the load constant is$(L_f)_2 = 2.763$ from column 7 of Table 15.5. Solving Eq. 15.10 gives $T_2 = 195$ million load traversals and, hence, friction has extended the endurance of the stud shear connectors from 90 million, in Example 15.3, to 195 million traversals and, hence, this section of the beam will last a further 65 years. Applying the same analysis to the quarter-span design point gives $T_2 = 282$ million which is equivalent to a further 94 years, and to the support design points gives $T_2 = 2,304$ million traversals which is equivalent to a further 768 years. It can be seen that friction has an enormous beneficial effect at the supports.

15.5 Crack propagation approach

The generic fatigue equation of Eq. 14.16 is adapted in this section to the crack propagation approach for stud shear connectors, in both a design mode and an assessment mode. Because the crack propagation approach allows for the direct interaction between strength and endurance, it is ideally suited for assessment.

15.5.1 *Design mode*
15.5.1.1 Crack propagation fatigue design equations

Comparing the generic fatigue material properties of Eqs. 14.6 and 14.7 with the stud shear connection crack propagation material properties in Eqs. 15.7 and 15.6 and substituting the corresponding values into the generic fatigue equation of Eq. 14.16, gives the following crack propagation fatigue equation.

$$\left(Q_{of} = f(n)\right)^{-5.1} = \frac{\left(1 - \dfrac{q_o}{\left(Q_{of} = f(n)\right)}\right)10^{3.12 - \frac{0.70}{\sqrt{n}}}}{\sum_{y=1}^{y=j}\left(T\left(F_f = f(q_t)\right)L_f\right)_y} = \frac{\left(1 - \dfrac{q_o}{\left(Q_{of} = f(n)\right)}\right)10^{3.12 - \frac{0.70}{\sqrt{n}}}}{T_1\left(F_f\right)_1\left(L_f\right)_1 + T_2\left(F_f\right)_2\left(L_f\right)_2 + \dots} \quad (15.11)$$

where q_o is the shear flow force due to the maximum overload, and Q_{of} is the shear flow strength of the stud shear connection when the structure is first built and which is required to resist not only the maximum overload q_o but also the reduction in strength due to fatigue loads. The function $F_f = f(q_t)$ in Eq. 15.11 is to remind the reader that the force spectrum must be derived in terms of shear flow forces, and the function $Q_{of} = f(n)$ is to remind the reader that the number of connectors required depends on their characteristic strength and, hence, the number n that can be assumed to fail as a group.

Equation 15.11 can be written in the following design form that is based on the mean material properties.

$$
Q_{of} = \left[\frac{1318\left(1 - \dfrac{q_o}{Q_{of}}\right)}{T_1\left(F_f\right)_1\left(L_f\right)_1 + T_2\left(F_f\right)_2\left(L_f\right)_2 + \ldots + T_j\left(F_f\right)_j\left(L_f\right)_j} \right]^{-\frac{1}{5.1}}
\tag{15.12}
$$

The aim of the design procedure is to determine the shear flow strength of the stud shear connectors when the structure is first built Q_{of}. It can be seen in Eqs. 15.11 and 15.12 that the parameter Q_{of} occurs on both sides of the equation, hence, an iterative approach has to be used.

It is worth noting that in general one standard fatigue vehicle with its associated load spectrum and force spectrum is used to represent all the fatigue vehicles, and therefore there is usually only one fatigue zone (TFL)$_1$ in Eq. 15.12. However, if two standard fatigue vehicles are required to represent all the fatigue vehicles, then the second standard fatigue vehicle and its associated values of T_2, $(L_f)_2$ and $(F_f)_2$ would form a second fatigue zone in Eq. 15.12. Similarly, if the same standard fatigue vehicle was moved across different lanes that crossed the bridge, then a fatigue zone would be created for the traversal across each lane.

15.5.1.2 **Example 15.5** *Fatigue design based on mean properties*
Let us apply Eq. 15.12 to the design of the composite bridge beam in Figure 15.1 in which T = 300 million. At the support design point: $F_f = 163 \times 10^9$ from column 3 of the no friction part of Table 15.2; $L_f = 1.14$ from column 4 in Table 15.1; $q_o = 1418$ N/mm from column 2 of the no friction part of Table 15.3. Hence TFL = 5.575×10^{19} and, therefore, Eq. 15.12 becomes

$$
Q_{of} = 1820\left(1 - \frac{\left(q_o = 1418\right)}{Q_{of}}\right)^{-0.1961}
\tag{15.13}
$$

Equation 15.13 has to be solved iteratively. For example, as we know that Q_{of} must be greater than $q_o = 1418$ N/mm, we could start with $Q_{of} = 2000$ N/mm which will make the right hand side equal to 2318 N/mm, which could then be used as the second estimate of Q_{of} for insertion into Eq. 15.13 in the next cycle of the calculation. The solutions converge fairly rapidly to $Q_{of} = 2220$ N/mm; the calculations being achieved readily on a spreadsheet program.

At the quarter-span design point: $F_f = 73 \times 10^9$ (Table 15.2); $L_f = 1.14$ (Table 15.1); $q_o = 1013$ N/mm (Table 15.3); $Q_{of} = 1555(1 - 1013/Q_{of})^{-0.196}$; which gives $Q_{of} = 1820$ N/mm. Finally at the mid-span design point: $F_f = 73 \times 10^9$; $L_f = 1.14$; $q_o = 608$ N/mm; $Q_{of} = 1555(1 - 608/Q_{of})^{-0.196}$; which gives $Q_{of} = 1700$ N/mm.

The shear flow strengths are plotted in Figure 15.4 as Q_{of}. The difference between Q_{of} and Q_o, that is shown hatched, is the reduction in strength due to the fatigue damage during the design life of the structure. The strength when the structure is first built has to be at least Q_{of}, so that at the end of the design life, the strength will not have reduced to less than Q_o, that is just sufficient to resist the maximum overload. Hence, Q_{of} will always be larger than the static requirement of Q_o, unless fatigue loads are not applied to the bridge in which case $Q_{of} = Q_o$.

At the supports TFL $= 5.57 \times 10^{19}$ and at mid-span TFL $= 2.50 \times 10^{19}$, therefore, the greater fatigue damage occurs at the supports. However, it can be seen in Figure 15.4 that the increase in strength due to fatigue damage $Q_{of} - Q_o$ is less at the supports than at mid-span. This is because the fatigue damage term TFL depends on the cyclic range q_t, whereas, the actual damage depends on the cyclic range as a proportion of the static strength, that is q_t/Q_{of}. Therefore, the increase in strength $Q_{of} - Q_o$ that is required at the supports is less than that at mid-span because the static strength requirement is larger at the supports.

15.5.1.4 **Example 15.6** *Varying the static strength requirement*

Let us consider the effect of varying the static strength requirement q_o of the composite bridge beam whilst maintaining the same fatigue damage, that is the same value of TFL. For example, it may be required to add more connectors to counteract the effects of creep, shrinkage or thermal gradients, or reduce slip at the concrete-slab/ steel-beam interface.

Applying Eq. 15.13 for different values of q_o at the supports gives the variation in Figure 15.5, where the ordinate is the proportional increase in strength required for fatigue. The variation in the proportional increase is asymptotic to the line $Q_{of}/q_o = 1$, because when $q_o \gg q_t$ the fatigue damage $(q_t/Q_{of})^{5.1} \to 0$. This means that if the static strength requirement is very large then the additional strength of connectors required for fatigue will be relatively small.

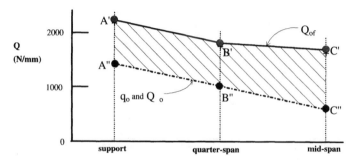

Figure 15.4 Shear flow strengths from crack propagation approach

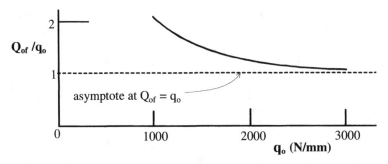

Figure 15.5 Variation of the static strength requirement

Another way of visualizing the effect of increasing the minimum static strength requirement is to double the static strength requirement in Example 15.5. Equation 15.13, that applies to the support, now becomes $Q_{of} = 1820(1 - 2836/Q_{of})^{-0.196}$ which gives $Q_{of} = 3050$ N/mm at the support. At the quarter-span $Q_{of} = 1555(1 - 2026/Q_{of})^{-0.196} = 2325$ N/mm and at mid span $Q_{of} = 1555(1 - 21216/Q_{of})^{-0.196} = 1900$ N/mm.

The results are plotted as the lines labelled $2Q_o$ and $Q_{of} = f(2Q_o)$ in Figure 15.6. Hence, the shaded region within points B is the increase in the shear connection strength that is required because of fatigue loads when the static strength is doubled. The partly hatched region within points A is the increase in strength for the original static strength from Figure 15.4. It can be seen that doubling the minimum static strength requirement has considerably reduced the additional connectors required for fatigue damage.

15.5.2 *Assessment mode*
15.5.2.1 Crack propagation fatigue assessment equations
The crack propagation design equation of Eq. 15.11 can be written in the following form that is suitable for assessment.

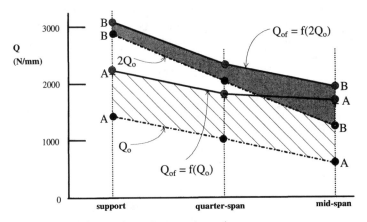

Figure 15.6 Doubling the static strength requirement

$$Q_{st}^{-5.1} = \frac{\left(1-\dfrac{Q_{res}}{Q_{st}}\right)10^{3.12-\frac{0.70}{\sqrt{n}}}}{\displaystyle\sum_{y=1}^{y=j}\left(TF_f\, L_f\right)_y} = \frac{\left(1-\dfrac{Q_{res}}{Q_{st}}\right)10^{3.12-\frac{0.70}{\sqrt{n}}}}{T_1\left(F_f\right)_1\left(L_f\right)_1 +\ldots\ldots+T_j\left(F_f\right)_j\left(L_f\right)_j} \quad (15.14)$$

where Q_{st} is the shear flow strength of the stud shear connectors prior to cyclic loading that can be taken as the strength when the structure was first built Q_{of}, Q_{res} is the residual or remaining strength after cyclic loading, and TFL are the fatigue zones that have or will occur. For assessment purposes, the unknown in Eq. 15.14 is either the residual strength Q_{res} or the residual endurance T_j after cyclic loading.

Assuming $n \to \infty$ in Eq. 15.14, gives the following assessment equation for deriving the mean residual strength

$$Q_{res} = Q_{st}\left[1 - \frac{Q_{st}^{-5.1}\sum TFL}{1318}\right] \quad (5.15)$$

that can be rearranged in the following form in order to determine the mean residual endurance T_j.

$$T_j = \frac{1318 Q_{st}^{5.1}\left(1-\dfrac{Q_{res}}{Q_{st}}\right) - \displaystyle\sum_{y=1}^{y=j-1}(TFL)_y}{F_j L_j} \quad (15.16)$$

It can be seen in the assessment equations of Eqs. 15.15 and 15.6, that the residual strength Q_{res} varies linearly with the fatigue damage parameter TFL. Therefore, the iterative solution required for the design of new structures in Section 15.5.1 is no longer required in this assessment mode.

15.5.2.2 **Example 15.7** *Assessing the remaining strength and endurance*

Let us now assess the performance of the composite bridge beam that was designed in Example 15.5. We will ignore the beneficial effect of friction and assess the performance throughout the whole design life at the mid-span design point. The original design parameters were: $T = 300$ million fatigue vehicle traversals during a design life of 100 years; $Q_o = 608$ N/mm from Table 15.3; $Q_{of} = 1700$ N/mm from Example 15.5; $F_f = 73 \times 10^9$ from Table 15.2; and $L_f = 1.14$ from row Table 15.1.

(a) Assessment equations

As $Q_{st} = Q_{of} = 1700$ N/mm, Eqs. 15.15 and 15.16 can be written as

$$Q_{res} = 1700 - (4.318 \times 10^{-17} \times \Sigma TFL) \quad (15.17)$$

and

$$T_j = \left(3.937 \times 10^{19} \left(1 - \left(Q_{res} / 1700\right)\right) - \sum TFL\right) / F_j L_j \qquad (15.18)$$

where the units again are N and mm.

(b) Initial reduction in strength

Let us assume that the number of vehicle traversals during the first 30 years is $T_1 = 150 \times 10^6$ that far exceeds the original anticipated rate of vehicle traversals of 300 million over 100 years. However, the distribution of the fatigue vehicle weights is as expected in the original design, so that the load constant remains unchanged at $(L_f)_1 = 1.14$. Furthermore, as no structural changes have been made to the bridge beam, the force constant also remains unchanged at $(F_f)_1 = 73 \times 10^9$.

The first fatigue zone is, therefore, given by $(TFL)_1 = 150 \times 10^6 \times 73 \times 10^9 \times 1.14 = 12.48 \times 10^{18}$. Inserting this value into Eq. 15.17 and bearing in mind that there is only one fatigue zone at this stage, gives $Q_{res} = 1154$ N/mm which is shown as point B in Figure 15.7. An alternative way of determining the residual strength is to remember that 150 million vehicle traversals have been applied of the original design number of 300 million. Therefore, as the residual strength varies linearly in a fatigue zone, the reduction in strength is given by $(Q_{st} - Q_o) \times (150/300) = 546$ N/mm and, hence, the residual strength is $1700 - 546 = 1154$ N/mm.

(c) Increasing the allowable commercial vehicle weights

We will now assume that the maximum weight of commercial vehicles has been allowed to increase. It is anticipated that level 4 in the load spectrum in column 2 in Table 15.1 will increase rapidly by 50%, so that the new load spectrum is given by columns 5 to 6 where the load constant has now increased to $(L_f)_2 = 2.10$.

The effect of the change in the commercial vehicle weights on the remaining endurance can be derived from Eq. 15.18. It is now required to determine the endurance when: the residual strength reduces to the minimum strength required to resist the maximum overload, that is $Q_{res} = Q_o = 608$ N/mm; $\sum TFL = (TFL)_1 = 12.48 \times 10^{18}$; $(L_f)_2 = 2.10$; and the force constant remains unchanged at $(F_f)_2 = (F_f)_1 = 73 \times 10^9$. Inserting these values into Eq. 15.18 gives $T_2 = 84 \times 10^6$ fatigue vehicle traversals which is shown as point D in Figure 15.7 and, hence, the increase in the weights of the commercial vehicles has reduced the remaining design life from a further 150 million traversal to 84 million traversals.

Let us now assume that 60 million of these fatigue vehicle traversals have been allowed to occur, so that the second fatigue zone $(TFL)_2 = 60 \times 10^6 \times 73 \times 10^9 \times 2.1 = 9.20 \times 10^{18}$ and, hence, $\sum TFL = (TFL)_1 + (TFL)_2 = 12.48 \times 10^{18} + 9.20 \times 10^{18} = 21.68 \times 10^{18}$. Applying Eq. 15.17 gives $Q_{res} = 762 \times 10^6$ which is shown as point C in Figure 15.7. Figure 15.7 clearly shows how the rate of the residual strength reduction due to the increase in the commercial vehicle weights has increased from fatigue zones 1 to 2.

(d) Placement of a weight restriction

Because of the rapid reduction in strength induced by the increase in the commercial vehicle weights, let us now assume that a weight restriction has to be placed on the bridge that eliminates levels 1 and 2 in Table 15.1 as shown in column 8. This will change the distribution of the probabilities as they have to be increased in proportion so that they sum to 1, as shown in column 9. Furthermore, the load constant in column 10 is now $(L_f)_3 = 1.46$.

Applying Eq. 15.18 with $(F_f)_3 = 73 \times 10^9$, $Q_{res} = Q_o = 608$ N/mm and $\Sigma TFL = 21.68 \times 10^{18}$, gives $T_3 = 34 \times 10^6$ which is the number of cycles required for the strength to reduce to the minimum requirement. This is shown as point E in Figure 15.7 and occurs after a total of 244 million fatigue vehicle traversals. If the rate of fatigue vehicle traversals remains constant at 150 million every 30 years, as occurred in the first fatigue zone, then the bridge will last a total of $(244/150) \times 30 = 49$ years.

(e) Remedial measures

At this stage, the strength of the stud shear connectors is at the minimum requirement of the strength to resist the maximum overload. The only way to increase the design life without reducing the maximum overload requirement, is to strengthen the bridge so that the shear flow forces are reduced. Let us assume that a plate is welded to the bottom flange of the composite beam, and that the addition of this plate reduces the original cross-sectional property of $A_c y_c/I_{nc}$ in Eq. 14.4 by 20%. Hence, the addition of the plate reduces K in Example 14.5 to 0.4×10^{-3} mm^{-1} and, therefore, reduces the shear flows to 80% of their original values prior to plating.

Reducing the shear flows to 80% of the original value, reduces the shear flows with no friction q_t in Table 15.2 by the same factor. Hence the force constant reduces by a factor of $0.8^{5.1} = 0.320$, so that at mid-span $(F_f)_4 = 0.320 \times 73 \times 10^9 = 23 \times 10^9$. Furthermore, the shear flow due to the maximum overload also reduces by 20% to $Q_o = 0.8 \times 608 = 486$ N/mm. The fatigue damage in the third fatigue zone that was previously analysed is $(TFL)_3 = 34 \times 10^6 \times 73 \times 10^9 \times 1.46 = 3.61 \times 10^{18}$, so that the total fatigue damage of the first three zones is now $\Sigma TFL = 25.29 \times 10^{18}$. Applying Eq. 15.18 with $Q_{res} = Q_o = 486$ N/mm, $\Sigma TFL = 25.29 \times 10^{18}$, $(F_f)_4 = 23 \times 10^9$, and with the value of the load factor prior to the weight restriction $(L_f)_4 = 2.10$, gives $T_4 = 58 \times 10^6$ which is shown as point F in Figure 15.7. The remedial work has, therefore, extended the life of the bridge to a total of 302 million traversals, that at the present rate of vehicle traversals will give a total design life of 60 years.

15.5.2.3 **Example 15.8** *Assessing the beneficial effect of friction*

The beneficial effect of friction is illustrated in this example by redoing the analyses in Example 15.7 with the effect of friction. It was shown in Example 15.4 that friction had the least effect on the endurance at the mid-span design point, so the analysis of this region will demonstrate the least benefit of friction on this beam. The effect of friction is to reduce the force constant from $(F_f) = 73 \times 10^9$ in column 9 in the no friction part of Table 15.2 to $(F_f) = 48 \times 10^9$ in column 9 in the with friction part.

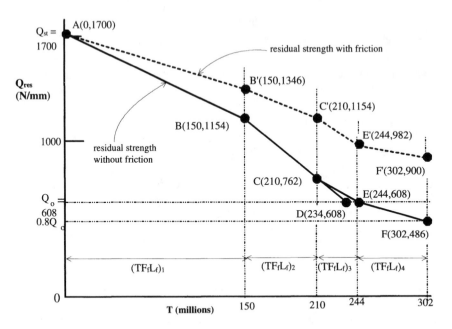

Figure 15.7 Variation in the residual strength

At zone 1 in Figure 15.7, $(TFL)_1 = 150 \times 10^6 \times 48 \times 10^9 \times 1.14 = 8.21 \times 10^{18}$, hence from Eq. 15.17, $Q_{res} = 1700 - (8.21 \times 10^{18} \times 4.318 \times 10^{-17}) = 1346$ N/mm as shown at point B' in Figure 15.5. At zone 2, $(TFL)_2 = 60 \times 10^6 \times 48 \times 10^9 \times 2.10 = 6.05 \times 10^{18}$, therefore, $\Sigma TFL = 8.21 \times 10^{18} + 6.05 \times 10^{18} = 14.26 \times 10^{18}$, hence, $Q_{res} = 1700 - (14.26 \times 10^{18} \times 4.318 \times 10^{-17}) = 1084$ N/mm as shown at point C'. At zone 3, $(TFL)_3 = 34 \times 10^6 \times 48 \times 10^9 \times 1.46 = 2.38 \times 10^{18}$, therefore, $\Sigma TFL = 16.63 \times 10^{18}$, hence, $Q_{res} = 1700 - (16.63 \times 10^{18} \times 4.318 \times 10^{-17}) = 982$ N/mm as shown at point E'. Finally at zone 4, $(TFL)_4 = 58 \times 10^6 \times 0.32 \times 48 \times 10^9 \times 2.10 = 1.89 \times 10^{18}$, therefore, $\Sigma TFL = 18.52 \times 10^{18}$, hence, $Q_{res} = 1700 - (18.52 \times 10^{18} \times 4.318 \times 10^{-17}) = 900$ N/mm as shown at point F'. It can be seen in Figure 15.7 that friction can substantially increase the residual strength and that friction provides a factor of safety against failure.

15.6 Composite building beam

15.6.1 *General*

Composite building beams can be subjected to fatigue loads such as those imposed by the traversal of cranes or fork lift trucks.

15.6.1.1 **Example 15.9** *Assessment of a composite beam in a building*

Let us consider the effect of the continuous applications of loads from the traversal of a fork lift truck as shown in Figure 15.8(a). It will be assumed that the fork lift truck does not cross the beam longitudinally but transversely and in line with the mid-span of the beam, so that the distance from the wheel loads V to the supports remains equal. It will also be assumed that the width between wheels remains constant at 1.6 m.

The composite beam spans 10 m and the material and geometric properties are given in Figure 4.3. The beam has already been designed in Example 4.1 in Section 4.2.2.2 using ultimate strength rigid plastic theory. In these analyses, the beam was designed with full shear connection, it had a flexural capacity of $M_{fsc} = 702$ kNm, and the strength of the shear connection in a half span was 2300 kN. Hence, the shear flow strength Q_{st} = 2300 × 10^3/5000 = 460 N/mm which is shown in Figure 15.8(b).

The elastic properties of the beam are given in Figure 5.4 in Section 5.3.5.2, where it was assumed that the beam was subjected to a long term load of w_{lng} = 20 kN/m. It will be assumed in this analysis that the only short term loads are the fork lift trucks. It will also be assumed that propped construction was used, so that all of the uniformly distributed dead load is resisted by the long term properties of the beam, and that all of the fork lift truck loads are resisted by the short term properties. From Figure 5.5, the long term value of K in Eq. 14.4 is $K_{lng} = (A_c y_c / I_{nc})$ = (28270 × 62)/716 × 10^6 = 2.49 × 10^{-3} and the short term value is $K_{lng} = (A_c y_c / I_{nc})$ = (57450 × 35)/819 × 10^6 = 2.46 × 10^{-3}.

(a) Minimum shear flow strength requirement

Assume that the beam must be able to resist a maximum overload that consists of a fork lift truck of 97 kN. This truck would induce a shear load of 48.5 kN and, hence, a short term shear flow force of 48,500 × 2.46 × 10^{-3} = 119 N/mm. The long term uniformly distributed load of 20 kN/m induces a maximum shear load of 100 kN at a support and, hence, a maximum long term shear flow force of 100 × 10^3 × 2.49 × 10^{-3} = 249 N/mm. Therefore, the maximum overload shear flow force q_o which is also the minimum strength requirement Q_o = 119 + 249 = 368 kN and occurs at the support as shown in Figure 15.8(b).

(b) Formation of a force and load spectrum

Let us assume that the weights W_{FV} and frequencies f of the fork lift trucks were measured over a day and the results are listed in columns 2 and 3 in Table 15.6. These weights have been converted to shear flow forces using the short term properties of the beam and are listed in column 4. Because of the loading arrangement, q_t is both the peak and range of the cyclic shear flow force induced by the lateral traversal of a vehicle.

A simple way to visualize the problem is to assume that columns 3 and 4 in Table 15.6 are the ranges and frequencies induced by a standard fatigue vehicle that has taken one day to traverse the beam. In which case, the force constant can be calculated as F_f = 1.974 x 10^{12} in column 5. The load constant L_f = 1, as only one vehicle equal to the weight of the standard fatigue vehicle is assumed to have traversed the beam

Table 15.6 Force spectrum for building beam

(1) Level	(2) W_{FV} (kN)	(3) f (per day)	(4) q_t (N/mm)	(5) $fq_t^{5.1}$ ($\times 10^{12}$)	(6) Q_{fric} (N/mm)	(7) q_{do} (N/mm)	(8) $fq_{do}^{5.1}$ ($\times 10^{12}$)
			Without friction		**With friction**		
1	80	100	98.2	1.445	4.8	93.4	1.119
2	50	400	61.4	0.527	3.0	58.4	0.408
3	10	300	12.3	0.002	0.6	11.7	0.000
			$F_f = 1.974 \times 10^{12}$			$F_f = 1.527 \times 10^{12}$	

and, hence, B = 1 and W = 1 in the load spectrum in Table 14.1, that will now only have one level. Furthermore T is now equal to the number of days, as it took the standard fatigue vehicle one day to cross the beam.

An alternative procedure for acquiring the force spectrum would have been to strain gauge the beam and record the variations in the strains for a fixed period. Columns 3 and 4 could have been derived from the magnitude and variations of the strain readings and T would now be the length of the period during which the strains were recorded.

(c) Endurance

Applying Eq. 15.16 in which $Q_{st} = 460$ N/mm, $Q_{res} = Q_o = 368$ N/mm, $F_f = 1.974 \times 10^{12}$ and $L_f = 1$ gives T = 5080 days. Hence after 5080 days, the shear flow strength of all

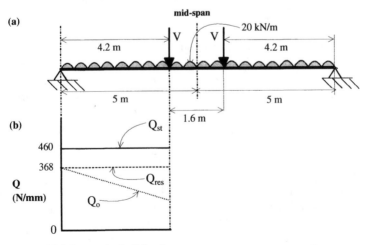

Figure 15.8 Composite building beam

the connectors in the shear span has reduced to $Q_{res} = 368$ N/mm as shown in Figure 15.8(b). The strength of the shear connectors at the support will be just sufficient to resist the maximum overload. However in the rest of the shear span there is a reserve of strength of $Q_{res} - Q_o$ as can be seen in Figure 15.8(b). Therefore, this endurance can be considered to be a lower bound.

(d) Residual flexural strength

As the cyclic shear flow forces are the same throughout the beam, because of the loading configuration shown in Figure 15.8(a), the strengths of all the connectors in a shear span reduce by the same amount, so that they will all have the same residual strength of 368 N/mm. Hence, the shear flow strength has reduced from 460 to 368 N/mm, that is the shear flow strength is $368/460 = 80\%$ of the strength when first constructed. As the beam was originally designed with full shear connection, the degree of shear connection is now $\eta_{max} = 0.8$. It was shown in Example 4.6 in Section 4.2.4.2 that this beam with a degree of shear connection of 80% has an ultimate rigid plastic strength of 653 kNm. Therefore, the fatigue loads will have reduced the flexural capacity of the composite from 720 kNm to 653 kNm in 5080 days.

(e) Beneficial effect of friction

Equation 14.5 has been used to determine the effect of friction. It has been assumed in the analyses that $\mu = 0.5$. The shear flow frictional resistances are given in column 6 in Table 15.6. This reduces the shear flows acting on the stud shear connectors from those in column 4 to those in column 7, from which $F_f = 1.527 \times 10^{12}$. Applying Eq. 15.15 in which $Q_{st} = 460$ N/mm, $T = 5080$, $F_f = 1.527 \times 10^{12}$ and $L_f = 1$, gives $Q_{res} = 389$ N/mm. Hence, the degree of shear connection is now $389/460 = 85\%$ which means that friction has increased the degree of shear connection by 5%.

15.7 References

1. Oehlers, D. J. (1990). 'Methods of estimating the fatigue endurances of stud shear connections.' International Association of Bridge and Structural Engineers. Proceedings P-145/90, Aug.
2. Oehlers, D.J. and Bradford, M.A. (1995). Composite Steel and Concrete Structural Members: Fundamental Behaviour. Pergamon Press, Oxford.

Index

Accumulated damage 211-12,214,234,237
Aggregate interlock 97,99,103
Amplification factor 132

Beams
    box 68
    composite
        efficient 58-59
        encased 67-70
        fatigue degradation 252-255
        non-standard 53
        standard 53-67,87-90
        strengthening 57,65.83,87,90
    hybrid 58
    L- 62,103,179
    profiled 4
Bearing
    zone 75,81
Bolting 2
Bond strength 102
Bracing 19
Bridges 34, 237-39
    fatigue degradation 235-255
Buckling 18-20
    flexural 121,125,131
    lateral-distortional
    19,20,53,190,197,199-208
        design models 203-208
        inelastic 207
        steel beams 200-203
    lateral-torsional 19,190,199
    local 18,20,27-38,53,69
        inelastic 186
        shear 28,29,35-36
    post-local 20
    strut model 203

Capacity
    flexural 53,57,59,61,70,71,130,134-139
        ducted beam 137-139,144,156-159
        effect of shear 66
        variation 70-73,84
    shear
        ducted beam 141-150,157-159
        reduction factor 6,
Centroid 51
Characteristic values 5
Component 1
Composite action 57,95
    double 3
Composite columns 2
    concrete-filled tube 2,4,36-38,107,108
    design recommendations 130-134
    encased 2,4,37,107,108

slender 3,107,121-134
stocky 3,107-120,121
Compression member 1
Concrete 8-10
    compressive strength 8,40,53,75
    crushing 9,12
    cube 8,53
    cylinder 8,53
    density 9,77
    effective modulus 10,40,46
    high-strength 7,9
    lightweight 80
    modulus of elasticity 8
    reinforced 11
    short-term properties 8-9,40
    stress-strain relationship 9
    tensile strength 11,26,54
        direct 9
        flexural 9
        lateral 9
        splitting 9
    thermal expansion
        coefficient 10
    wet 2,48
Confinement
    post-cracking concrete 180-182
Construction
    propped 2,48-9,51,83,225,238
    unpropped 2,48,225,238
Continuous beams 23,44-45,185-198
Contraflexure 23,45
Cover 60,162,168
Cracking
    initiation 209,235-237
        approach 214,239-245
    longitudinal 97,177
    propagation 209,235,237
        approach 212,214,245-252
Creep 45-46,235
    coefficient 10,45,48
    strain 10
Cyclic loading 218-227,231,235

Deflections 23
Deformation capacity 11
Deformations 49,79
    instantaneous 45
    time-dependent 45,47,48
Design
    action 6
    point 70,212,228,237,247
        shear flow 212,228
Dimensions
    effective 168

Dispersal 162,165
Dowel
    action 95,97,222
    force 95
    resistance 77-84
    stiffness 39
    strength 51,77-98
        maximum 12
        post-cracking 177-184
        tensile 80
Ductility 11,53,69,80,185,188
    parameter 194-195

Effective
    depth 102
    length 125-127
    prism 165-166
    section 25-26
    sizes 23
    width 21-26,55,57
Elastic analysis 39-52,77,83-84,121-27,206
Elastic design 11
Endurance 209,235,242,254
    based fatigue properties 209-210,235
    characteristic 209-211
Equilibrium
    of forces 12,53-55,70
Equivalent material strength 66-67
Equivalent prism 172-176
    effective sides 173-176
Equivalent stress system 63
Euler load 121-122,124

Failure
    embedment 135,159-161
    envelope 93,112,150-151
        composite 84
        fracture 93-94
    splitting 162-176
Fatigue
    analysis 212-214
        stud shear connections 235-255
    damage 209,212,214,218,232-233,235,247
    design 5
    equation 231-234,239,243,245,248
        generic analysis procedures 209-234
    load
        frequency 217
        zone 234,244
Fatigue vehicles 238
    standard 215,216,218,232
Finite element method 207
First yield analysis 42-44,185
Flexural rigidity 43-44
Force constant 233,234
Forces 6
    cyclic range 222-225,231
    distribution of 55,60

embedment 159-161
    flexural
        distribution 53
Force spectrum 218,239,254
Force transfer 1
Formwork 3,4,37
Fracture
    excessive slip 84-94
    strength analysis 84-94
Frequency distribution 5
Friction
    coefficient of 227,238
    force at interface 226
    passive 97
    resistance 227-231,245,251,255

Gluing 4

Haunch 25-26,60,80,86,105-106
Hogging region (*see* Negative region)

Imperfection parameter 123-125
Influence line diagram 218-226,228-231
Instability (*see* Buckling)
Interaction
    analysis 40-47
    degree of 17-18
    full 17,40,83,185
    no 17,47
    partial 13,17-18,48
    shear and flexure 150-151
Interface 12,57
Internal support 18
Interpolation
    linear 72
Inverted U-frame 203
    design approach 204-208
Iterative solution 79,82,143,243,246

Limit states design 4-6
    serviceability 6,11,39
    ultimate 7,209
Load
    bridge 214-218
    concentrated 64,162,182
    constant 233,234,238
    dead 6
    distributed 67,88,91
    dowel 53
    live 6
    long-term 45,48,83,93
    nominal 6
    non-uniform 82
    patch 162
    point 89
    service 10
    short-term 40,48,83,93,107
    spectrum 216-218,239,253
    traversals 64,214-217

Load-Slip path 75-79

Material properties 7-12
    linear 39-40
    rigid plastic 10-12
    variation 80
Mean strength 5
Mixed Analysis Approach 86,88,93
Modified slenderness 28-37,216
Modular ratio 40
Moment
    primary 127
    secondary 127-130
    local 143
    maximum 53
Moment-curvature relationship
    115,187,195

Negative bending 19,43,44,104,200
Negative region 19,27,44,45,67,199
Neutral axis
    factor 11,110-114
    position 30
    32,35,41,53,56,57,61,62,67,88,114
        in concrete component 16,43,56
        in steel component 16
Nominal action 6
Nonlinearity 107,114-115

Overstress 51

Parametric Study Approach 86-88
Perimeter length 95,99,102,105
Plastic
    centroid 108-110
    design (*see* Rigid plastic analysis)
    hinge 186-197
    moment
        hogging 194
        sagging 194-195
        rotation 188-190
Plating 154-156
Poisson's ratio 8
Positive bending 22,28,35,40,44
Positive region 23,44,45
Prediction equation 5
Propped cantilever 190-191
Push tests 74,77

Radius of gyration 123,208
Range of load 209
Redistribution
    bending moments 185-197
        continuous beams 190-194
        steel beams 186-193
Regression analysis 209
Reservoir method 224
Residual strength 209,211-214,235,255
Restraint

    triaxial 75
Reinforcing
    bars 9
        hooped 104,177-180
    deformed bars 8
    fabric 8
    mesh 8
    transverse 86,97-104
        anchorage 101-104
Ribs
    longitudinal 60,101
    oblique 26
    profiled 62
    transverse 26,60,62,64,80-82,137
Rigid plastic
    analysis 11,19,34,39,53-73,77,115-
    120,135,185,200
    properties 10-12,196
Rotation
    beam 188,190
    requirements 193

Sagging region (*see* Positive region)
Second order effects 121,122
Section
    classification 28
    35,42,108,190,197,206
    compact 28,31-32
    equivalent 61
    plastic 28-29
    semi-compact 28,32-33
    slender 28,37
Separation
    vertical 1
Service ducts 67,135-161
Shape factor 185
Shear
    lag 21-27
    transfer 141-143
Shear connection
    degree of 13,15,47,64,67,70,84,138
    full 14,17,55,137
    no 57
    partial 12-16,47-48,55,57,64
    shape 76
Shear connectors
    characteristic strength 78
    configurations 76,78,81,168-172
    deformations 40
    detailing 75-77,78
    distribution 52,70,72,76
    77,82,83,90-93,96,242
        uniform 70
    fracture 53,66,78
    mechanical 1,74-93
    spacing 76-77
    stud (*see also* Dowel) 1,3
        diameter 74
        head 1,74

height 74,76
 shank 1,77,79
 weld collar 1,74,76,80
 welded 232
 stiffness 39,40
Shear flow 21
 forces 74,95-97,222,235
  envelope 213,228
 on connectors 51-52,235,238
 planes 95-106
 strength 54,70,83,214,235,242-242,247,253
  generic 97-98
  mean 72
Shear force 77
 longitudinal 95
 transfer 95-106
Shear plane
 encompassing shear connector 104-106
 through depth of slab 99
Shear span 54,63,78
Sheeting
 profiled 3,4,25,80,100-102
 thickness 3
Shrinkage 9,235
 strains 9,48
Slabs
 profiled 60,100,137
 shear resistance 151-154
Slenderness 202-203,207
 limits 28-38
 parameter 124,132-134
 ratio 124-125
Slip 16,40,74,75,85
 distribution 16
 excessive 70-75
 maximum 86-88,90
 strain 16,17,40
Splitting 58,77
 design philosophy 58
 global 164
 local 162-176
 longitudinal 97
 plane 179
 post 64,182-184
 strength 164-168
 zone 162
Squash load 122-124
Standard deviation 5,211
Statically indeterminate 5,185
Steel
 column 121-127
 elastic modulus 7,40
 fracture 11
 mild 7
 yielding 7,40,54
Steel decking 8
Stiffness 12
Strain distribution 12,110-111,139

Strain hardening 7,11,186,195
Strains
 elastic 10
 instantaneous 9
 yield 7
Strength 13
 characteristic of dowel 51
 concrete component 54
 design 6
 mean of dowel 77,80,84
 shear connection 54-57
 shear connectors 12,14,57,74,168-172
 steel component 33,56,55
 ultimate 11
Stress
 biaxial 7
 flexural 23,49-50
 proof 7
 resultants 13
 shear yield 7
 tensile 209
 uniaxial yield 6
Stress distributions 12,34,64,68

Transformed area 40-44,83,109
Transverse flexure 99,104-105
Transverse reinforcement 100-106,177
Triaxial confinement 177-179

Ultimate strength
 analysis 39
Understress 51

Vehicles
 traversal 209
Voids 62,80,102

Warping displacements 21
Webs
 openings 135-161

Yield
 moment 33,185,186

Printed in the United States
by Baker & Taylor Publisher Services